Student Activities and Lab Manual to Accompany Basic Robotics

Keith Dinwiddie

CENGAGE
Learning·

Australia • Brazil • Mexico • Singapore • United Kingdom • United States

CENGAGE
Learning·

Student Activities and Lab Manual to Accompany Basic Robotics
Keith Dinwiddie

SVP, GM Skills & Global Product Management:
Dawn Gerrain

Product Team Manager: Erin Brennan

Associate Product Manager: Nicole Sgueglia

Senior Director, Development:
Marah Bellegarde

Senior Product Development Manager:
Larry Main

Senior Content Developer: Sharon Chambliss

Product Assistant: Jason Koumourdas

Vice President, Marketing Services:
Jennifer Ann Baker

Marketing Manager: Kelsey L. Hagan

Senior Production Director: Wendy Troeger

Production Director: Andrew Crouth

Content Production Management
and Art Direction: Lumina Datamatics, Inc.

Cover image(s): Robotic Arm: ©Alexey
Dudoladov/iStock/Thinkstock; Two robotic
hand tools: ©Ociacia/iStock/Thinkstock;
Aldebaran Robotics: © AFP/Getty Images;
Dextre, the Canadian Space Agency's robotic
handyman: © Stocktrek Images/Getty Images

For product information and technology assistance, contact us at
Cengage Learning Customer & Sales Support, 1-800-354-9706

For permission to use material from this text or product,
submit all requests online at **www.cengage.com/permissions**.
Further permissions questions can be e-mailed to
permissionrequest@cengage.com

Library of Congress Control Number: 2014943299

ISBN: 978-1-285-42278-7

Cengage Learning
20 Channel Center Street
Boston, MA 02210
USA

Cengage Learning is a leading provider of customized learning solutions with office locations around the globe, including Singapore, the United Kingdom, Australia, Mexico, Brazil, and Japan. Locate your local office at:
www.cengage.com/global

Cengage Learning products are represented in Canada by Nelson Education, Ltd.

To learn more about Cengage Learning, visit **www.cengage.com**

Purchase any of our products at your local college store or at our preferred online store **www.cengagebrain.com**

Notice to the Reader

Publisher does not warrant or guarantee any of the products described herein or perform any independent analysis in connection with any of the product information contained herein. Publisher does not assume, and expressly disclaims, any obligation to obtain and include information other than that provided to it by the manufacturer. The reader is expressly warned to consider and adopt all safety precautions that might be indicated by the activities described herein and to avoid all potential hazards. By following the instructions contained herein, the reader willingly assumes all risks in connection with such instructions. The publisher makes no representations or warranties of any kind, including but not limited to, the warranties of fitness for particular purpose or merchantability, nor are any such representations implied with respect to the material set forth herein, and the publisher takes no responsibility with respect to such material. The publisher shall not be liable for any special, consequential, or exemplary damages resulting, in whole or part, from the readers' use of, or reliance upon, this material.

Printed in the United States of America
Print Number: 01   Print Year: 2014

Table of Contents

Preface

About This Activities Manual

Activity manuals are common in the world of textbooks, but I think if you take a few moments to explore this one in particular, you will find some nice surprises. After students have learned about the material in the textbook, this activities manual will give them a chance to test and see how much stuck while discovering the areas that need more attention on their part. For the definitions, there is a matching section. For the general material, multiple-choice and short answer questions focus on the key points of the textbook. For the chapters in which mathematical material is covered, there are practice problems so that students can get used to working with the equations.

After students have completed the above work, they will find an essay and research section for each chapter. The essay section will challenge students to write a paper based on what they have learned from the accompanying textbook chapter. For some of these assignments, students will relate what they learned to the classroom lab equipment, while other assignments will require students to demonstrate understanding and help them pull it all together. The research section is where students get to search for current information outside of the textbook, related to the topics covered, and create a report. Given how fast the world of robotics is changing and growing, this is a great way to make sure students keep up with current events in the field. A separate rubric is provided for the essay section and the research section so that students know exactly what is expected.

Each chapter finishes up with a suggested lab that fits well with the topics covered in the text. Some of the suggested labs correspond with labs provided in the activities manual, while others utilize the classroom lab equipment in a specific way. Along with the suggested lab, there is a lab form for the student to fill out and turn in that describes the lab, details the steps taken, provides a place for observations, and finishes with a conclusion section where the student can express what he or she has discovered and learned. For labs other than those suggested, blank lab forms are at the back of the manual, and the template is in the instructor's materials. In the instructor's materials is also a grading template for the labs that can be modified as needed. The labs provided in the manual are low-cost labs that offer students a chance to get hands-on experience working with the various types of components used in robotic systems. If not used in the classroom, these labs provide a great avenue for those students who wish to learn more and work with robots at home, as the materials are easy to acquire.

The whole purpose of all these activities is to provide a range of options for learning. Need to embed some English in your course? Use the essay assignments. Students complaining about what the book covers? Have them do the reports and find their own information. Have a student who is excelling in the classroom labs? Challenge him or her with the Spider Bot lab. Having trouble getting all the material fitted in as is but want the students to get some extra practice? Assign just the core matching, multiple-choice, problem, and short answer questions. The beauty of all the options provided in this activities manual is that the material gives instructors a variety of ways to meet the objectives of their specific course.

Activities

About This Section

In this section of the manual, you will find activities that correspond to the chapters of the *Basic Robotics* textbook. Each chapter's activities include multiple-choice, matching, and short answer questions along with math practice problems as appropriate. In addition to these standard activities, each chapter has an essay and research assignment as well as a suggested lab. You should check with your instructor to confirm if he or she wishes for you to complete all the activities or only a portion.

If you are completing a lab other than the suggested lab, you can find blank forms at the back of the manual. Again, check with your instructor for specific direction on reporting lab activities as well as points possible and grading guidelines. Make sure to follow all the safety rules of your classroom and to keep your wits about you as you complete labs. Labs are often the more exciting part of a robotics class, but they are also the most dangerous. Remember the three Rs of robotics: Robots Require Respect. It is a good idea to review Chapter 2 on safety from time to time, especially as you get into the more complex labs.

The instructions for the robotic build labs are in Section II of the activities manual. Each lab details the materials needed, lists the steps involved, and has pictures to help guide you through the lab. It is always best to look through the lab first and make sure you have everything you need before you actually begin.

CHAPTER 1 | History of Robotics

Name: _____ Date: _____

Score: _____ Text pages 1–39

ACTIVITIES

Multiple Choice (1 point each)

Identify the choice that best completes the statement or answers the question.

_____ 1. Who launched and docked a robotic capsule with the international space station in 2012?
   a. SpaceX
   b. NASA
   c. Russia
   d. FANUC

_____ 2. The Baxter robot was created by which of the following companies?
   a. ReWalk
   b. Rethink Robotics
   c. Aldebaran
   d. Honda

_____ 3. A job that involves work with a very dusty substance would fall into which of the following categories?
   a. dull
   b. dirty
   c. difficult
   d. dangerous

_____ 4. Who designed the Stanford Arm?
   a. AMF
   b. Richard Hohn
   c. David Silver
   d. Victor Scheinman

_____ 5. "A robot may not injure a human being or, through inaction, allow a human being to come to harm" was written by _____.
   a. Isaac Asimov
   b. Dr. Mark Tilden
   c. Karel Capek
   d. Fritz Lang

_____ 6. Which of the following is a benefit to surgeons who use robotic surgical systems?

     a. increased concentration

     b. working outside the operating room

     c. two doctors working together

     d. all of these

_____ 7. Which company released the T3 robot arm?

     a. Cincinnati Milacron

     b. AMF

     c. Unimate

     d. KUKA

_____ 8. Which company shipped a VERSATRAN robot to Japan for industrial use?

     a. Unimate

     b. AMF

     c. Kawasaki Robotics

     d. KUKA

_____ 9. Who built Herbert Televox for Westinghouse Electric and Manufacturing Co.?

     a. Joseph Barnet

     b. Fritz Lang

     c. Nikola Tesla

     d. Roy Wensley

_____ 10. Who is credited with writing the first example of a computer program?

     a. Thomas Edison

     b. Augusta Ada King

     c. Charles Babbage

     d. Herman Hollerith

_____ 11. The Robosapien is the brainchild of _____.

     a. Dr. Mark Tilden

     b. Dr. Hod Lipson

     c. Ralph Hollis

     d. Cynthia Breazeal

_____ 12. Who created the Lute Player Lady?

     a. William Oughtred

     b. Lorenz Rosenegge

     c. Gianello Torriano

     d. Friedrich von Knauss

_____ 13. Working in an area of high radiation would fall into what category?

     a. dull

     b. dirty

     c. difficult

     d. dangerous

_____ **14.** Which company holds the honor of inventing the first robot controller capable of synchronizing two robots?

    a. FANUC

    b. ABB

    c. MOTOMAN

    d. Unimate

_____ **15.** Who produced the IR 600 system in 1978?

    a. Unimate

    b. AMF

    c. Kawasaki Robotics

    d. KUKA

_____ **16.** The film *Metropolis* was the work of _____.

    a. Fritz Lang

    b. Karel Capek

    c. Alan Turing

    d. Marvin Minsky

_____ **17.** Who created the ASIMO robot?

    a. Honda

    b. GM

    c. NASA

    d. FANUC

_____ **18.** If loading a machine required the operator to bend and twist at an awkward angle, what category would this task fall under?

    a. dull

    b. dirty

    c. difficult

    d. dangerous

_____ **19.** Who built Electro the Moto-Man for Westinghouse Electric and Manufacturing Co.?

    a. Joseph Barnet

    b. Roy Wensley

    c. Harold Roselund

    d. Thomas H. Flowers

_____ **20.** When did the da Vinci medical robot receive approval for surgery?

    a. 2000

    b. 1998

    c. 1994

    d. 1999

_____ **21.** The NAO robot belongs to which of the following companies?

    a. Microsoft

    b. WowWee

    c. MOTOMAN

    d. Aldebaran

_____ **22.** Loading the same part in the same way multiple times a day would fall into what category?

      a. dull

      b. dirty

      c. difficult

      d. dangerous

_____ **23.** ASEA became what modern company?

      a. FANUC

      b. ABB

      c. MOTOMAN

      d. PUMA

_____ **24.** Which of the following companies developed a robot with a human torso for industry?

      a. FANUC

      b. ABB

      c. MOTOMAN

      d. AMF

_____ **25.** Who shipped the first commercial NC machine?

      a. FANUC

      b. ABB

      c. MOTOMAN

      d. Unimate

Matching (1 point each)

Match the terms to the definitions below.

a.	Abacus	**n.**	Fringe benefits
b.	Algorithm	**o.**	Fuzzy logic
c.	Anthropomorphic	**p.**	Gantry
d.	Automata	**q.**	NC
e.	Automatons	**r.**	Paraplegic
f.	BEAM	**s.**	Precision
g.	Bionic	**t.**	Prosthetic
h.	Biosensor	**u.**	Punch card control system
i.	Character recognition	**v.**	Repeatability
j.	CNC	**w.**	Robot
k.	Consistency	**x.**	Robota
l.	EOAT	**y.**	Show-stoppers
m.	Exoskeleton	**z.**	Telepresence

_____ **26.** Health and dental insurance, retirement, life insurance, social security, and anything else that the company pays part or all of the cost of for the employee.

_____ **27.** A counting device that uses beads to help the user keep track of numbers and make calculations; one of the first attempts of our ancestors to conquer complex math.

_____ **28.** Computer numerically controlled.

_____ **29.** Someone who has lost all mobility in the lower limbs.

_____ **30.** Device that picks up the electrical activity associated with movement thru wires incorporated into the surface of the patient's skin as well as needle-sized electrodes implanted directly into the muscles.

_____ **31.** Drudgery or slave-like labor.

_____ **32.** Performing tasks accurately or exactly within given quality guidelines.

_____ **33.** The ability to produce the same results or quality each time.

_____ **34.** To be like a human.

_____ **35.** The ability to perform the same motions within a set tolerance.

_____ **36.** The ability to read written or printed letters, numbers, and symbols.

_____ **37.** A machine equipped with various data gathering devices, processing equipment, and tools for operational flexibility and interaction with the systems environment, which is capable of carrying out complex auctions under either programmed control or direct manual control.

_____ **38.** Name given to many simple, two-axis or three-axis machines designed to pick up parts from one area and place them in another.

_____ **39.** Biology Electronics Aesthetics Mechanics.

_____ **40.** Situation in which a person performs interactions with others from a remote location using technology.

_____ **41.** Devices built for the sole purpose of getting people to stop and take notice.

_____ **42.** Artificial replacement part for the body that closely resembles the missing part in appearance and function.

_____ **43.** Devices that work under their own power; often designed to mimic people.

_____ **44.** The part of the robot that manipulates parts or performs tasks.

_____ **45.** Self-operated machines.

_____ **46.** System that works by using various holes in a metal or wooden rectangle called a "card" to control the timing and flow of actions.

_____ **47.** Robotic systems that strap onto and around the user's body to enhance strength, endurance, and, in some cases, mobility.

_____ **48.** Systematic procedure for solving a problem or accomplishing a task.

_____ **49.** Robotic system designed to mimic a human system and controlled by nerve impulses.

_____ **50.** Numerically controlled.

_____ **51.** Situations in which there is more than one right answer and one must come up with a plan of action.

Matching (1 point each)

Match the answers to the questions below.

a.	Archytas of Tarentum	**j.**	Campbell Aird
b.	Gottfried Wilhelm	**k.**	Hans Bullmann
c.	John Dee	**l.**	Herman Hollerith
d.	Rechenuhr	**m.**	Ktesibios of Alexandria
e.	10–17 C.E.	**n.**	Jack Kilby
f.	Pierre Jaquet-Droz	**o.**	Karel Capek
g.	Sir William Gove	**p.**	Leonardo da Vinci
h.	Charles Babbage	**q.**	Joseph John Thomson
i.	Thomas H. Flowers	**r.**	Gorge Boole

_____ **52.** Who was fitted with the world's first fully bionic arm?

_____ **53.** Who gave the world the word "robot"?

_____ **54.** Who ended up charged with witchcraft for his mechanical creation?

_____ **55.** Who created the Writer?

_____ **56.** Who created the Metal-Plated Warrior?

_____ **57.** Who is credited as the father of pneumatics?

_____ **58.** Who created Boolean algebra?

_____ **59.** Who holds the honor of creating the first true robot?

_____ **60.** Who created the world's first fuel cell in 1839?

_____ **61.** Who developed the binary system of arithmetic?

_____ **62.** Who created the analytical engine with processor and memory that many credit as the world's first computer?

_____ **63.** Who created a wooden pigeon that could fly with steam or compressed air?

_____ **64.** Who created the company that would ultimately become IBM?

_____ **65.** Who inferred the existence and characteristics of the electron?

_____ **66.** Who created the microchip?

_____ **67.** The first mechanical calculator that could add, subtract, multiply, and divide was known as the _____.

_____ **68.** Who lead the group that created Colossus, which is known as the world's first electronic computer?

_____ **69.** When was the first steam engine design created?

Short Answer (2 points each)

Write the answers to the following questions in the space provided.

70. What does ASIMO stand for, and when did Honda introduce the system?

_____

_____

_____

_____

_____

_____

71. What was PLANETBOT, and what is special about it?

_____

_____

_____

_____

_____

_____

_____

72. What type of environment negates the need for any AI?

_____
_____
_____
_____
_____
_____

73. How was Electro the Moto-Man controlled?

_____
_____
_____
_____
_____
_____

74. What is unique about the robot created at Cornell University by Josh Bongard, Victor Zykoy, and Hod Lipson?

_____
_____
_____
_____
_____
_____

75. What does AESOP stand for?

_____
_____
_____
_____
_____

76. Who gave the world the word "robot," and how did they introduce it?

_____
_____
_____
_____
_____

77. What is the driving force behind the use of telepresence systems?

_____

_____

_____

_____

_____

_____

78. What are the three laws of BEAM robotics?

_____

_____

_____

_____

_____

_____

79. What is the benefit of direct drive robotic systems, and when where they first introduced?

_____

_____

_____

_____

_____

_____

_____

Name: _____   Date: _____

Essay Assignment (12 points)

80. Review the time line of events from the beginning of Chapter 1 and pick the event that you feel was the most influential in the development of the modern robot. Be sure to detail why you picked this event, why you think it is more important than the other events, and how you feel this event influenced the modern robots of today. Make sure you fully express your ideas on the subject so that anyone who reads your answer can understand your point of view. Use the Rubric and any directions from your instructor to guide your work.

RUBRIC	1	2	3	4	Points Earned
CONTENT	The work contains only a chosen event and is missing the other three key points of why the student chose it, why it was more important than the rest, and how it influenced the modern robot.	The work is missing two of the key points of why the student chose it, why it was more important than the rest, or how it influenced the modern robot.	The work is missing one of the key points of why the student chose it, why it was more important than the rest, or how it influenced the modern robot.	The work chooses an event, details why the student chose it, why it was more important than the rest, and how it influenced the modern robot.	
ORGANIZATION	There is no organization of the material.	There is some organization of material, but it is still difficult to follow.	The work has a clear structure, but some of the organization interferes with clarity.	The work is clear and concise.	
GRAMMAR	There are four or more spelling, punctuation, or other grammar mistakes.	There are two or three spelling, punctuation, or other grammar mistakes.	There is one spelling, punctuation, or other grammar mistake.	Spelling, punctuation, and grammar are all correct.	

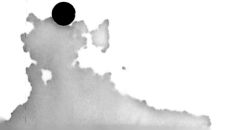

Name: _____ Date: _____

Research Assignment (12 points)

81. What is a robot?

Ask five people outside of your class what a robot is and record the answers given. Once you have these answers compare them to the various definitions for a robot given in Chapter 1 and see which fits best for each. Finish up with any trends you notice in the answers or any modifications you feel should be made to the definition of a robot.

RUBRIC	1	2	3	4	Points Earned
CONTENT	The work has less than five researched answers listed, no indication of which definition fits, and is missing the trend observation or definition modification.	The work has less than five researched answers listed along with an indication of which definition fits and is missing the trend observation or definition modification.	The work has all five researched answers listed along with an indication of which definition fits, but it is missing the trend observation or definition modification.	The work has all five researched answers listed along with an indication of which definition fits and the trend observation or definition modification.	
ORGANIZATION	There is no organization of the material.	There is some organization of material, but it is still difficult to follow.	The work has a clear structure, but some of the organization interferes with clarity.	The work is clear and concise.	
GRAMMAR	There are four or more spelling, punctuation, or other grammar mistakes.	There are two or three spelling, punctuation, or other grammar mistakes.	There is one spelling, punctuation, or other grammar mistake.	Spelling, punctuation, and grammar are all correct.	

Suggested Lab (value assigned by instructor)

82. To begin your lab experience the recommended lab for Chapter 1 is the Vibro Bot lab from the lab section of the activities manual. In this lab you will take some simple components and create an autonomous system ready to go off and explore the world. This robot falls into the BEAM category, as it is a simple system with no sensors and a great way to get started in the world of robotics. I also encourage you to build several of these robots and change something on each one to see how it changes the behavior of the system. As a class you may want to get all the groups' robots together and see what happens when you have several in an enclosed space. This is the essence of research into swarm robotics, something several universities are studying as of the writing of this manual.

Use the provided lab form to assist with reporting your lab activities, unless directed otherwise by your instructor.

Vibro Bot Lab Form

Lab Description

In the space below, describe the purpose of the lab and the equipment involved.

_____
_____
_____
_____
_____
_____
_____
_____
_____
_____
_____
_____

Lab Execution

In the space provided, detail the steps you took to perform the lab. Make sure to include any troubleshooting steps performed.

_____
_____
_____
_____
_____
_____
_____
_____
_____
_____
_____

Observations

Record your observations about the system's performance here, including both the expected and unexpected.

_____

_____

_____

_____

_____

_____

_____

_____

_____

_____

Conclusions

What conclusions or statements can you make about the robot based on your observations and any data gathered during the course of the lab?

_____

_____

_____

_____

_____

_____

_____

_____

_____

_____

Name: _____ Date: _____

Score: _____ Text pages 40–61

ACTIVITIES

Multiple Choice (1 point each)

Identify the choice that best completes the statement or answers the question.

_____ 1. Which of the following is **not** one of the three R's of robotics?

    a. robots

    b. respect

    c. recycle

    d. require

_____ 2. The area where anyone could travel around the robot would be the _____.

    a. work envelope

    b. safe zone

    c. cautionary zone

    d. danger zone

_____ 3. The area where you are close to the robot but still out of its reach would be the _____.

    a. work envelope

    b. safe zone

    c. cautionary zone

    d. danger zone

_____ 4. Another name for the work envelope would be the _____.

    a. work envelope

    b. safe zone

    c. cautionary zone

    d. danger zone

_____ 5. The area where the operator often performs his or her tasks around the robot would be the _____.

    a. work envelope

    b. safe zone

    c. cautionary zone

    d. danger zone

_____ 6. The area where you are required by OSHA to have the teach pendant would be the _____.

    a. danger zone

    b. safe zone

    c. cautionary zone

    d. none of these

_____ 7. The common amperage ratting for most 110-V circuits, like wall outlets, is usually _____.

    a. 30 A

    b. 10 A

    c. 15 A

    d. 100 A

_____ 8. A common path for electricity through the body is _____.

    a. finger to different hand

    b. finger to same hand

    c. ear to ear

    d. hand to foot

_____ 9. Which of the following is NOT part of a worst-case scenario for a shock victim?

    a. The victim loses muscular control.

    b. There is not enough current flowing to trip breakers or fuses.

    c. The victim is part of the circuit.

    d. Someone knocks the victim free of the circuit within seconds of the shock using a wooden two-by-four.

_____ 10. General rule number one of emergencies is _____.

    a. remain calm

    b. perform to the level of your training

    c. assess the situation

    d. after it is over, talk it out

_____ 11. General rule number two of emergencies is _____.

    a. remain calm

    b. perform to the level of your training

    c. assess the situation

    d. after it is over, talk it out

_____ 12. General rule number three of emergencies is _____.

    a. remain calm

    b. perform to the level of your training

    c. assess the situation

    d. after it is over, talk it out

_____ 13. General rule number four of emergencies is _____.

    a. remain calm

    b. perform to the level of your training

    c. assess the situation

    d. after it is over, talk it out

_____ 14. Which of the following is true about tourniquets?

    a. They are a last resort.

    b. They can damage the victim's tissue.

    c. They should be used only by trained personnel.

    d. All of these.

_____ **15.** If the initial bandage on a severe cut is soaking through with blood, you should _____.

    a. place a clean bandage on top of the soiled one

    b. lower the wound below the heart if possible

    c. remove it and replace it

    d. apply a tourniquet

_____ **16.** Burns that heal in about a week fall into the category of _____ burns.

    a. first-degree

    b. second-degree

    c. third-degree

    d. fourth-degree

_____ **17.** A red burn with blistering and a wet look is a _____ burn.

    a. first-degree

    b. second-degree

    c. third-degree

    d. fourth-degree

_____ **18.** A burn that reaches the bone is a _____ burn.

    a. first-degree

    b. second-degree

    c. third-degree

    d. fourth-degree

_____ **19.** A burn that leaves leathery skin and an open wound is a _____ burn.

    a. first-degree

    b. second-degree

    c. third-degree

    d. fourth-degree

_____ **20.** A burn that causes some loss of sensation in the affected area is a _____ burn.

    a. first-degree

    b. second-degree

    c. third-degree

    d. fourth-degree

_____ **21.** Which of the following would we use to treat a minor burn?

    a. Submerge the area in cool water for 10–15 minutes.

    b. Pop any blisters that appear.

    c. For persistent pain treat with over-the-counter pain killers for one week.

    d. All of these.

_____ **22.** For a fourth-degree burn, which of the following would we do?

    a. Place in cool water.

    b. Remove any clothing stuck to the burn.

    c. Seek immediate medical help.

    d. All of these.

_____ **23.** The number one enemy when dealing with burns is _____.

    a. pain

    b. blisters

    c. time

    d. infection

_____ **24.** Which of the following would NOT make a good splint?
       a. two pieces of wood
       b. two metal rods
       c. rolled up magazines
       d. none of these

_____ **25.** If someone is being electrocuted and you are not able to turn off the power, which of the following would NOT be a good choice for freeing the victim?
       a. a wooden two-by-four
       b. a wooden broom handle
       c. a dry nonconductive rope
       d. kicking them

_____ **26.** Which of the following describes the action of the robot after pushing an E-stop?
       a. halts motion as quickly as possible
       b. shuts down many of the power and drives systems
       c. creates an alarm condition preventing normal operation
       d. all of these

Numeric Response (2 points each)

Use the formulas from the textbook to answer the following questions.

27. How much electrical force does it take to push 10 amps of current past 10 ohms of resistance?

_____

_____

_____

_____

28. What is the amperage of a system with 50 Ω of resistance and a voltage of 120 V?

_____

_____

_____

29. What is the resistance of a circuit that is powered by 120 V and has 2 A of current flowing through it?

_____

_____

_____

_____

30. What voltage is required to push 20 A past 24 Ω?

_____

_____

_____

_____

31. There is an industrial robot that has 240 V coming in and is pulling 30 A on a measured line. What is the total resistance for this circuit?

_____

_____

_____

_____

32. How many amps are in a circuit feed by 220 V with a resistance of 15.5 Ω?

_____

_____

_____

_____

Matching (1 point each)

Match the amperage to the effects on the human body.

a.	0.001 A to 0.003 A	**e.**	0.100 A to 0.200 A
b.	0.003 A to 0.010 A	**f.**	0.200 A to 0.300 A
c.	0.010 A to 0.030 A	**g.**	2.0 A to 4.0 A
d.	0.030 A to 0.100 A		

_____ 33. Muscle contractions and breathing difficulty begins with loss of muscle control possible

_____ 34. Ventricular fibrillation highly possible

_____ 35. Heart stops beating, internal organ damage occurs, irreversible bodily damage possible

_____ 36. Ranges from unnoticed to mild sensation

_____ 37. Severe shock with high possibility of respiratory paralysis

_____ 38. Severe burns and breathing stops

_____ 39. Painful shock

Matching (1 point each)

Match the resistance below to the material from Chart 2-2 in the book.

a.	200,000 Ω to 200,000,000 Ω	**e.**	as low as 1,000 Ω
b.	2,000 Ω to 100,000 Ω	**f.**	as low as 150 Ω
c.	1 Ω	**g.**	400 Ω to 600 Ω
d.	100,000 Ω to 600,000 Ω	**h.**	100 Ω

_____ 40. Dry skin

_____ 41. Hand to foot (inside body)

_____ 42. Wet wood 1 inch thick

_____ 43. Damp skin

_____ 44. 1,000 feet of 10 AWG copper wire

_____ 45. Ear to ear (inside body)

_____ 46. Wet skin

_____ 47. Dry wood 1 inch thick

Matching (1 point each)

Match the terms below with the appropriate definitions.

a.	Amperage	**n.**	Photo eye sensor
b.	Blunt force trauma	**o.**	Presence sensors
c.	Cautionary zone	**p.**	Pressure sensor
d.	Circuit	**q.**	Proximity switch
e.	Danger zone	**r.**	Resistance
f.	Electricity	**s.**	Safe zone
g.	Emergency	**t.**	Safety interlock
h.	E-stop	**u.**	Shock
i.	Expanded metal guarding	**v.**	Teach pendant
j.	Guards	**w.**	Tourniquet
k.	Grounded point	**x.**	Voltage
l.	Light curtain	**y.**	Ventricular fibrillation
m.	Limit switch	**z.**	Work envelope

_____ **48.** The area that the robot can reach during operation

_____ **49.** A system in which all the safety switches have to be closed or made for the equipment to run in automatic

_____ **50.** The area where one is close to the robot, but still outside of the work envelope or reach of the system

_____ **51.** Devices designed to protect us from the dangers of a system

_____ **52.** A device that generates an electromagnetic field and senses the presence of various materials by changes in this field with no physical contact with the item sensed

_____ **53.** A measurement of how many electrons, or how much electricity, is flowing through a system

_____ **54.** Metal that is perforated and stretched to create diamond-shaped holes with quarter-inch pieces of metal around it

_____ **55.** Sensors that detect when a person is inside the danger zone and that are tied into the system to prevent automatic operation

_____ **56.** Emits an infrared beam that is reflected back by a shiny surface or a standard reflector to a receiver in the unit

_____ **57.** A set of circumstances or a situation that requires immediate action and often involves the potential for or events that have caused injury to people and/or severe damage to property

_____ **58.** Detects the presence or absence of a predetermined or set level of force

_____ **59.** This is a measurement of how much force is working against the flow of electrons

_____ **60.** The area where a person can pass near the robot without having to worry about making contact with the system

_____ **61.** A path that electrons flow through

_____ **62.** A tightened band that restricts arterial blood flow to wounds of the arms or legs to stop severe bleeding

_____ **63.** A sensor that houses the emitter and receiver separately to create an infrared sensing barrier

_____ **64.** This force drives electricity through a system

_____ **65.** An impact that does not penetrate the skin

_____ **66.** A handheld device, usually attached to a fairly long cord that allows people to edit or create programs, control various operations of the robot

_____ **67.** A point somehow connected to the earth

_____ **68.** Emergency stop

_____ **69.** Senses the presence or absence of a material by contact with a movable arm attached to the end of the unit

_____ **70.** The area the robot can reach or the work envelop and where all the robot's tasks take place

_____ **71.** A flow of electrons from a point with more electrons to a place with fewer electrons

_____ **72.** A condition in which the heart quivers instead of actually pumping blood

_____ **73.** When a person becomes a part of the circuit

Short Answer (2 point each)

Write the answers to the following questions in the space provided.

74. What is one of the reasons robots seem to start up for no reason?

_____

_____

_____

_____

75. What are some of the alarm conditions that could stop a robot from operating?

_____

_____

_____

_____

76. What are some of the mechanical conditions that could stop a robot?

_____

_____

_____

_____

77. What do we use metal mesh for in relation to robot safety?

_____

_____

_____

_____

78. What does OSHA stand for?

_____

_____

_____

_____

79. What is the purpose behind having the teach pendant with you anytime you are in the robot's work envelope?

_____

_____

_____

_____

80. What is the purpose of guarding?

_____

_____

_____

81. How do metal cages protect people from the danger zone?

_____

_____

_____

82. How can we use cameras to guard the danger zone?

_____

_____

_____

83. What is the primary difference between a limit switch and a prox switch?

_____

_____

_____

84. Who invented Ohm's law?

_____

_____

_____

85. How long does 0.1 A of current need to pass through the heart to have the potential to cause ventricular fibrillation?

_____

_____

_____

86. Why is CPR often a large part of treating someone who is electrocuted?

_____

_____

_____

87. How do we treat minor impact injuries?

_____

_____

_____

Essay Assignment (12 points)

88. Review the information from this chapter on the three zones that surround the robot and then use this information to define each zone for the robotic equipment used in your classroom or lab environment. Make sure to include the distance from the robot for each zone, the shape of the zone, and any other important factors that you would need in order to explain these areas to someone entering your classroom for the first time. If you are not sure of the size of the work envelope for your robot, ask you instructor for direction on finding this information.

RUBRIC	1	2	3	4	Points Earned
CONTENT	There are problems with the area definitions as well as their dimensions.	At least two areas are clearly defined, and one or two are accurately dimensioned.	All three areas are clearly defined, but only two are accurately dimensioned.	All three areas are clearly defined and accurately dimensioned.	
ORGANIZATION	There is no organization of the material.	There is some organization of the material, but it is still difficult to follow.	The work has a clear structure, but some of the organization interferes with clarity.	The work is clear and concise.	
GRAMMAR	There are four or more spelling, punctuation, or other grammar mistakes.	There are two or three spelling, punctuation, or other grammar mistakes.	There is one spelling, punctuation, or other grammar mistake.	Spelling, punctuation, and grammar are all correct.	

Research Assignment (12 points)

89. Using the information from this chapter and other reputable resources such as OSHA, RIA, or NIOSH (National Institute for Occupational Safety and Health), create a list of 10 safety rules for your class related lab activities. Once you have your rules, make sure you list them in order of importance, with number 1 being the most important and 10 the least. Include with your list the reasons behind your ordering of the safety rules. Make sure your rules are clear enough that if they were posted on the wall, a person reading them for the first time would understand what to do.

RUBRIC	1	2	3	4	Points Earned
CONTENT	There are 10 rules but the list is missing ordering, justification, and clarity of what the rules require.	There are 10 rules, but the work is missing two of the following: placed in order of importance, written in such a way anyone could understand them, with the ordering justifications included.	There are 10 rules, placed in order of importance, but there is at least one that is unclear, with the ordering justifications included.	There are 10 rules, placed in order of importance, written in such a way anyone could understand them, with the ordering justifications included.	
ORGANIZATION	There is no organization of the material.	There is some organization of the material, but it is still difficult to follow.	The work has a clear structure, but some of the organization interferes with clarity.	The work is clear and concise.	
GRAMMAR	There are four or more spelling, punctuation, or other grammar mistakes.	There are two or three spelling, punctuation, or other grammar mistakes.	There is one spelling, punctuation, or other grammar mistake.	Spelling, punctuation, and grammar are all correct.	

Suggested Lab (value assigned by instructor)

90. Before you get too deep into the lab experience, you need to ensure you can work safely with the robot(s) in your lab. To this end, have your instructor walk you through the safety rules and requirements of the lab area and then perform any demonstrations as necessary. To augment this lab, take some time and analyze the robot's safety systems to determine if there is anything that needs to be added, modified, or repaired. Remember, the whole point of safety systems and requirements is to keep YOU safe, not the robot. You have a vested interest in making sure you can run the robot(s) safely as well as knowing how to shut things down should something go wrong.

Use the provided lab form to assist with reporting your lab activities, unless directed otherwise by your instructor.

Safety Lab Form

Lab Description

In the space below, describe the purpose of the lab and the equipment involved.

_____
_____
_____
_____
_____
_____
_____
_____
_____
_____
_____
_____

Lab Execution

In the space provided, detail the steps you took to perform the lab. Make sure to include any troubleshooting steps performed.

_____
_____
_____
_____
_____
_____
_____
_____
_____
_____
_____
_____

Observations

Record your observations about the systems performance here, including both the expected and unexpected.

_____
_____
_____
_____
_____
_____
_____
_____
_____
_____

Conclusions

What conclusions or statements can you make about the robot based on your observations and any data gathered during the course of the lab?

_____
_____
_____
_____
_____
_____
_____
_____
_____
_____

Name: _____ Date: _____

Score: _____ Text pages 62–87

ACTIVITIES

Multiple Choice (1 point each)

Identify the choice that best completes the statement or answers the question.

_____ **1.** Most three-phase systems have _____.
- **a.** three hot wires and a neutral
- **b.** three hot wires but no ground
- **c.** three hot wires, a neutral, and a ground
- **d.** three hot wires and a ground

_____ **2.** We compare the _____ of the robot to human wrist.
- **a.** controller
- **b.** major axes
- **c.** minor axes
- **d.** teach pendant

_____ **3.** The Robosapien and LEGO NXT run on _____.
- **a.** AC electricity
- **b.** DC electricity
- **c.** both of these
- **d.** neither of these

_____ **4.** Most industrial robots run on _____.
- **a.** AC electricity
- **b.** DC electricity
- **c.** both of these
- **d.** neither of these

_____ **5.** Which of the following axes in a six-axes robot would be considered the pitch or up and down orientation of the wrist?
- **a.** 3
- **b.** 4
- **c.** 5
- **d.** 6

_____ 6. _____ stays at a constant intensity.

    a. Single-phase AC

    b. Three-phase AC

    c. DC voltage

    d. Multiphase DC

_____ 7. The power in a standard 110-V wall outlet is _____.

    a. single-phase AC

    b. three-phase AC

    c. DC voltage

    d. multiphase DC

_____ 8. We consider the _____ to be the brain of the robot.

    a. controller

    b. major axes

    c. minor axes

    d. teach pendant

_____ 9. We use the RMS value of _____ to determine the EMF.

    a. AC electricity

    b. DC electricity

    c. both of these

    d. neither of these

_____ 10. The power in your home is _____.

    a. AC electricity

    b. DC electricity

    c. both of these

    d. neither of these

_____ 11. The _____ wire of a 110-V plug provides the return path of the system.

    a. hot

    b. neutral

    c. ground

    d. all of these

_____ 12. The electrons flow in only one direction in _____ systems.

    a. single-phase AC

    b. three-phase AC

    c. DC voltage

    d. multiphase DC

_____ 13. _____ has a set polarity.

    a. AC electricity

    b. DC electricity

    c. both of these

    d. neither of these

_____ **14.** The _____ wire of a 110-V plug provides the electricity to the system.

    a. hot

    b. neutral

    c. ground

    d. all of these

_____ **15.** The first fluid used in hydraulics was _____.

    a. olive oil

    b. crude oil

    c. whale oil

    d. water

_____ **16.** When using water in hydraulic systems, the user has to watch out for _____.

    a. rusting of ferrous metal parts

    b. freezing

    c. bacterial growth

    d. all of these

_____ **17.** The power in most businesses is _____.

    a. AC electricity

    b. DC electricity

    c. both of these

    d. neither of these

_____ **18.** Robots can run on _____.

    a. AC electricity

    b. DC electricity

    c. both of these

    d. neither of these

_____ **19.** We compare the _____ of the robot to the human torso and arm.

    a. controller

    b. major axes

    c. minor axes

    d. teach pendant

_____ **20.** Which of the following axes in a six-axes robot would be considered the yaw or side-to-side orientation of the wrist?

    a. 3

    b. 4

    c. 5

    d. 6

_____ **21.** _____ does not have a set polarity.

    a. AC electricity

    b. DC electricity

    c. both of these

    d. neither of these

_____ **22.** Battery and solar panels provide _____.

    a. AC electricity

    b. DC electricity

    c. both of these

    d. neither of these

_____ **23.** Pinhole hydraulic leaks _____.

    a. are the most dangerous type of hydraulic leak

    b. can turn hydraulic oil into a flammable mist

    c. can generate pressures that shear through human tissue

    d. all of these

_____ **24.** Using supplied _____ negates the concerns about amp-hours.

    a. AC electricity

    b. DC electricity

    c. both of these

    d. neither of these

_____ **25.** The dead man switch on the teach pendant _____.

    a. is used during manual control of the robot

    b. stops the robot if released when in use

    c. stops the robot if pressed too hard when in use

    d. all of these

_____ **26.** When hooking up batteries in series, if you place one of the batteries backward, or positive to positive, what happens to the system?

    a. Nothing.

    b. It doubles the output of the system.

    c. It subtracts from the system.

    d. None of these.

_____ **27.** _____ systems do not require a neutral wire due to the way the voltage flows.

    a. Single-phase AC

    b. Three-phase AC

    c. DC voltage

    d. Multiphase DC

_____ **28.** Which of the following axes in a six-axes robot would be considered the roll or rotation orientation of the wrist?

    a. 3

    b. 4

    c. 5

    d. 6

_____ **29.** The major axes of the robot are _____

    a. axes 1–3

    b. pitch, yaw, and roll

    c. axes 4–6

    d. all of these

_____ **30.** The _____ wire of a 110-V plug provides a low-resistance path for electrons to flow through when the insulation is broken or component failure occurs.

    a. hot

    b. neutral

    c. ground

    d. all of these

_____ **31.** _____ has instances when no voltage is present.

    a. Single-phase AC

    b. Three-phase AC

    c. DC voltage

    d. Multiphase DC

_____ **32.** Power that has three sine waves 120 degrees apart electrically is _____.

    a. single-phase AC

    b. three-phase AC

    c. DC voltage

    d. multiphase DC

Matching (1 point each)

Match the terms to the definitions below.

a. Alternating current/AC   **n.** Hydraulic power

b. Amp-hours/Ah   **o.** Major axes

c. Axis   **p.** Manipulator

d. Base   **q.** Minor axes

e. Controller   **r.** Mobile base

f. Cycle   **s.** Neutral wire

g. Direct current/DC   **t.** Normally closed/NC

h. Degree of freedom/DOF   **u.** Normally open/NO

i. Electromotive force/EMF   **v.** Relay logic

j. External axis   **w.** Single-phase AC

k. Gantry base   **x.** Solid mount base

l. Ground   **y.** Teach pendant

m. Hertz/Hz   **z.** Three-phase AC

_____ **33.** The use of a noncompressible liquid given velocity and then piped somewhere to do work

_____ **34.** Contacts that do pass power when the relay is de-energized

_____ **35.** Where we bolt the robot for stability with nonmobile types or the mobile platform on which a manipulator is mounted

_____ **36.** Voltage in which the electrons flow back and forth in the circuit

_____ **37.** Responsible for the orientation and positioning of the tool

_____ **38.** Responsible for getting whatever tool we are using into the general area it needs to be in

_____ **39.** Sine wave cycles per second

_____ **40.** One complete wave from zero to positive to zero to negative and back to zero

_____ **41.** AC power that has one sine wave provided to the system via a single hot wire and returned on a neutral wire

_____ **42.** For industrial systems, these are often robotic arms or overhead systems with a series of rods that move the tooling around

_____ **43.** The brains of the operation and the part of the robot responsible for executing actions in a specific order and timing or under specified conditions

_____ **44.** The preferred way to go or mounting the robot firmly to the floor or other structures using bolts and fastening systems

_____ **45.** Linear bases with a finite reach

_____ **46.** Axis or axes of motion that often move parts, position tooling for quick change, or in some other way help with the tasks of the robot

_____ **47.** Each part of the robot that has controlled movement

_____ **48.** Another name for the force that drives electrons through a circuit; is synonymous with voltage

_____ **49.** Control system that uses devices know as relays to create various logic-sorting situations, which would in turn control the operation of the system

_____ **50.** Contacts that do not allow power through when the relay is de-energized

_____ **51.** Number of amps a power source can deliver over a period of time

_____ **52.** Each axis of the robot that gives it one more way to move

_____ **53.** The wire that provides a return path for the electrons and what allows for a complete circuit

_____ **54.** Systems used to move the manipulator to various locations so that it can perform its functions

_____ **55.** Voltage in which the electrons flow in only one direction

_____ **56.** AC power that has three sine waves 120 degrees apart electrically

_____ **57.** Device that allows the operator to view alarms, make manual movements, stop the robot, change the program or start a new program, and any of the other day-to-day tasks required of those who run robots

_____ **58.** The wire that provides a low-resistance path for electrons to flow through when the electrons escape the system due to insulation or component failure

Matching (1 point each)

Match the terms to the definitions below.

a. Amperes/amps
b. Conventional current flow theory
c. Electricity
d. Electron flow theory
e. Pneumatic power
f. Polarity
g. Program
h. Relay
i. Resistance
j. Root Mean Square/RMS
k. Voltage

_____ **59.** A measurement of the potential difference or imbalance of electrons between two points and the force that will cause electrons to flow

_____ **60.** Devices that use a small control voltage to make or break connections between field devices

_____ **61.** The opposition to the flow of electrons in the circuit and the reason electrical systems generate heat during normal operation

_____ **62.** A system of logic filters and commands

_____ **63.** The flow of electrons from a place of excess to a place of deficit that we route through components to do work

_____ **64.** A mathematical average of the sine wave

_____ **65.** Electron flow from the positive terminal to the negative terminal

_____ **66.** The positive and negative terminals of components

_____ **67.** Electron flow from the negative terminal to the positive terminal

_____ **68.** A measurement of electrons passing a given point in one second

_____ **69.** A fluid power that uses air to generate force

Short Answer (2 points each)

Write the answers to the following questions in the space provided.

70. What are the three questions we use to determine which power source is best for a robot?

_____

_____

_____

_____

_____

_____

_____

_____

_____

_____

_____

71. When we measure 110 V at the wall outlet, what are we measuring?

_____

_____

_____

_____

_____

_____

_____

72. Why are linear mobile bases often called gantry bases?

_____

_____

_____

_____

_____

_____

73. How do we hook up batteries to increase the amp-hours, and how do we hook them up to increase voltage? (Be sure to include polarity in your answer.)

_____

_____

_____

_____

_____

_____

_____

74. What happened during World War II that changed the world of hydraulics?

_____

_____

_____

_____

_____

_____

_____

75. Why do we ground circuits?

_____

_____

_____

_____

_____

76. How do we hold position in pneumatic systems?

_____

_____

_____

_____

_____

_____

77. Using the image below, number the axes of the robot as instructed in the chapter. When you are done, there should be six axes labeled on your robot.

_____

_____

_____

_____

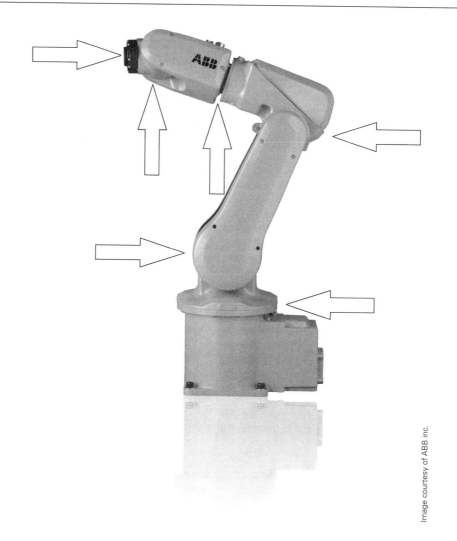

Image courtesy of ABB inc.

78. Is it possible to have a three-phase AC system that has no neutral wire? Fully support your answer as to why you said yes or no.

_____

_____

_____

_____

_____

79. What do we mean when we say that the United States uses 60-hertz power?

_____

_____

_____

_____

_____

Problem (2 points each)

Answer each of the following questions in the space provided.

80. Given three D cell batteries at 1.5 V each, draw a power system that would increase the total voltage. What would the voltage of this new system be?

81. If you have a 6-Ah battery, how long could you run a robot that needs 1.5 amps?

_____

_____

_____

_____

82. If you have a 10-Ah battery, how long could your run a robot that requires 0.5 amps?

_____

_____

_____

_____

_____

83. You are creating a robot in the classroom, and it is your job to create the power supply. The robot runs on 12-V DC at 2 amps, and you are tasked with making sure this robot can run for at least five hours on a set of batteries. For your portion of the project, you have been given a large supply of D cell batteries that are 1.5 V with 5 Ah of charge in them. Draw below how you would create this power supply for your robot.

84. Given three D cell batteries that each have 5 Ah of charge, draw below how you would connect those three batteries to increase the Ah of the power source. How many Ah total would this system have?

_____

_____

_____

_____

85. You are working to create an industrial-sized battery-powered robot to move materials around a facility. This system runs on 48 V and needs to be able to run eight hours between charges with the normal amperage draw of the system being 80 amps. You are using 12 V 100 Ah deep cycle batteries that resemble car batteries to build the power supply. Draw below how you would create this system. (Feel free to look up images of car batteries if you do not know what these look like.)

Essay Assignment (12 points)

86. Take what you learned from this chapter and detail the various parts of the robot you will be using in your class labs. Make sure you detail the type of power supply your robot uses, the type of controller, the type of teach pendant used (if present), what type of robot it is with DOF, and the type of base your robot uses. If you are using a kit like the LEGO NXT or VEX systems, tailor this information to a lab designated by your instructor. (If you are unsure which lab to use or you have multiple robots in your lab, please ask your instructor for direction.)

RUBRIC	1	2	3	4	Points Earned
CONTENT	The following parts for the directed robot are detailed, with at least one part correct: power supply, controller, teach pendant, robot type and DOF, robot base type.	The following parts for the directed robot are detailed, with two or three errors: power supply, controller, teach pendant, robot type and DOF, robot base type.	The following parts for the directed robot are detailed, with no more than one error: power supply, controller, teach pendant, robot type and DOF, robot base type.	The following parts for the directed robot are detailed: power supply, controller, teach pendant, robot type and DOF, robot base type.	
ORGANIZATION	There is no organization of the material.	There is some organization of the material, but it is still difficult to follow.	The work has a clear structure, but some of the organization interferes with clarity.	The work is clear and concise.	
GRAMMAR	There are four or more spelling, punctuation, or other grammar mistakes.	There are two or three spelling, punctuation, or other grammar mistakes.	There is one spelling, punctuation, or other grammar mistake.	Spelling, punctuation, and grammar are all correct.	

Research Assignment (12 points)

87. Using either material present in your classroom or reputable research sources, pick a complex system such as the Robonaut, Baxter, NAO, the Mars rover, or a specific industrial system and write a short paper detailing what robot you chose, the five major components of that robot, and the sources you used for information to write your report. Make sure you detail the type of power supply the robot uses, the type of controller, the type of teach pendant used (if present), what type of robot it is with DOF, and the type of base the robot uses. If using resources outside of the classroom, make sure you provide references (in the format preferred by your instructor) for said resources.

RUBRIC	1	2	3	4	Points Earned
CONTENT	The report has only two of the following: what robot you chose, the five parts of that robot, and all applicable references.	Multiple pieces of the following information are missing: what robot you chose, the five parts of that robot, and all applicable references with.	Only one piece of the following information is missing: what robot you chose, the five parts of that robot, and all applicable references.	The report details what robot you chose, the five parts of that robot, and all applicable references.	
ORGANIZATION	There is no organization of the material.	There is some organization of material, but it is still difficult to follow.	The work has a clear structure, but some of the organization interferes with clarity.	The work is clear and concise.	
GRAMMAR	There are four or more spelling, punctuation, or other grammar mistakes.	There are two or three spelling, punctuation, or other grammar mistakes.	There is one spelling, punctuation, or other grammar mistake.	Spelling, punctuation, and grammar are all correct.	

Suggested Lab (value assigned by instructor)

88. This is great time to explore the robots you will use in the lab for your course. With your instructor's approval, learn the following about the robot(s) you will work with for this course:

- How to power the robot up
- How to move the robot in manual
- The different modes of the robot
- All the safety features
- The different parts of the robot
- Anything else deemed important by your instructor

The point of this lab is to learn the basics of the robot(s) you will work with during this course so that you can start to build your robotic knowledge. I have had many students who could move the robot to a designated point in manual during the day we practiced movement, only to have a large amount of trouble performing the same task down the road during programming. That is why it is crucial you actually learn all you need to know about moving and working with the robot instead of just going through the motions. This will save you time and frustration later when you start programming and doing other complex labs.

Use the provided lab form to assist with reporting your lab activities, unless directed otherwise by your instructor.

Equipment Introduction Lab Form

Lab Description

In the space below, describe the purpose of the lab and the equipment involved.

_____
_____
_____
_____
_____
_____
_____
_____
_____
_____

Lab Execution

In the space provided, detail the steps you took to perform the lab. Make sure to include any troubleshooting steps performed.

_____
_____
_____
_____
_____
_____
_____
_____
_____

Observations

Record your observations about the system's performance here, including both the expected and unexpected.

_____

_____

_____

_____

_____

_____

_____

_____

_____

Conclusions

What conclusions or statements can you make about the robot based on your observations and any data gathered during the course of the lab?

_____

_____

_____

_____

_____

_____

_____

_____

_____

_____

CHAPTER 4 | Classification of Robots

Name: _____ Date: _____

Score: _____ Text pages 88–119

ACTIVITIES

Multiple Choice (1 point each)

Identify the choice that best completes the statement or answers the question.

_____ 1. To make sure the block travels straight along the ball screw, we use _____.

   a. an encoder
   b. a guide rod
   c. limit switches
   d. all of these

_____ 2. Air leaks can _____.

   a. send chips flying
   b. whip around severed lines
   c. create drops in system pressure
   d. all of these

_____ 3. A gear used to change direction is called the _____.

   a. drive gear
   b. driven gear
   c. idler gear
   d. compound gear

_____ 4. Which of the following geometries is the most common due to the flexibility of the design and the fact it can replicate a wide range of human motions?

   a. Cartesian
   b. cylindrical
   c. spherical
   d. articulate

_____ 5. Gear systems with _____ can have both drive and driven gears on the same shaft.

   a. an odd number of gears
   b. an even number of gears
   c. a compound gear
   d. an idler

_____ **6.** The ISO subcategories for personal care robots are _____.

     a. mobile servant robot

     b. physical assistant robot

     c. person carrier robot

     d. all of these

_____ **7.** The common feedback device for servomotors is the _____.

     a. photo eye

     b. proximity switch

     c. limit switch

     d. encoder

_____ **8.** Gear systems with _____ will have the output rotating the same as the input.

     a. an odd number of gears

     b. an even number of gears

     c. a compound gear

     d. an idler

_____ **9.** Worm gears _____.

     a. are simpler in construction than bevel gears

     b. can generate torques up to 500:1

     c. can be locked in place by applying enough pressure on the spur or helical gear

     d. all of these

_____ **10.** The downside to synchronous belts is _____.

     a. they slip easily

     b. they are often damaged by machine crashes

     c. they can run only in the clockwise direction

     d. none of these

_____ **11.** The gear tied to the motor or power supply is called the _____.

     a. drive gear

     b. driven gear

     c. idler gear

     d. compound gear

_____ **12.** Pneumatic power is often used in _____.

     a. robot positioning

     b. robot tooling

     c. heavy payload applications

     d. all of these

_____ **13.** To increase torque, we usually sacrifice _____.

     a. precision

     b. speed

     c. repeatability

     d. all of these

_____ **14.** Which of the following robots are often used in the electronics field, where their motion and strengths seem a good fit for the tasks required?

    a. articulated

    b. Cartesian

    c. SCARA

    d. pneumatic

_____ **15.** We call two gears on the same shaft a _____.

    a. drive gear

    b. driven gear

    c. idler gear

    d. compound gear

_____ **16.** Early robots and applications where there are other means of verifying position commonly use _____.

    a. stepper motors

    b. servo motors

    c. VFDs

    d. all of these

_____ **17.** For proper operation, the _____ and _____ of a gear should match.

    a. pitch diameter/number of teeth

    b. diametral pitch/pressure angle

    c. pressure angle/pitch diameter

    d. diametral pitch/pitch diameter

_____ **18.** Gear systems with _____ will have the output rotating opposite the input.

    a. an odd number of gears

    b. an even number of gears

    c. a compound gear

    d. an idler

_____ **19.** Which of the following geometries has the benefits of moving materials over large distances and saving floor space by mounting over the equipment?

    a. Cartesian

    b. cylindrical

    c. spherical

    d. articulate

_____ **20.** Hypoid gears can _____.

    a. generate gear ratios of 60:1 but are always louder than spiral bevel gears

    b. generate only low torque ratios

    c. generate more power than worm gears

    d. generate gear ratios of 60:1

_____ **21.** We typically change hydraulic oil _____.

    a. weekly

    b. monthly

    c. yearly

    d. only when something catastrophic happens to the system

_____ **22.** A bevel gear set can rotate the force between _____ degrees

    a. 0 and 45

    b. 0 and 90

    c. 0 and 270

    d. 0 and 180

_____ **23.** The gear tied to the output is called the _____.

    a. drive gear

    b. driven gear

    c. idler gear

    d. compound gear

_____ **24.** Which of the following geometries has the benefits of being good for reaching deep into machines, saving on floor space, and tending to have the rigid structure needed for large payloads?

    a. Cartesian

    b. cylindrical

    c. spherical

    d. articulate

_____ **25.** When a Cartesian robot has only two major axes, what is the common axis left out?

    a. X-axis

    b. Y-axis

    c. Z-axis

    d. C-axis

Matching (1 point each)

Match the terms to the definitions below.

a.	Spiral bevel gear	**m.**	Sprocket
b.	International Standards Organization	**n.**	Spur gears
c.	Mesh	**o.**	Stepper motors
d.	Miter gear	**p.**	Synchronous belt
e.	Pitch diameter	**q.**	Torque
f.	Pressure angle	**r.**	Transmission
g.	Reduction drive	**s.**	V-belt
h.	RPM	**t.**	Velocity
i.	SCARA	**u.**	Worm gears
j.	Servomotors	**v.**	Rack and pinion
k.	Skew gears	**w.**	Zerol bevel gears
l.	Slippage		

_____ **26.** The matching up of gear teeth

_____ **27.** A type of bevel gear with the tooth curved to reduce noise and smooth out operation

_____ **28.** Belts shaped like a V that are used to transfer force

_____ **29.** Bevel gears with equal numbers of teeth and the shafts at 90°

_____ **30.** How the forces interact between two gears, at what angles, and how a gear tooth is rounded or shaped

_____ **31.** A continuous rotation-type motor with built-in feedback devices called encoders

_____ **32.** Systems that take motor output and, via mechanical means, reduce or alter it

_____ **33.** Gears made by taking a round or cylindrical object and cutting teeth into the edge

_____ **34.** Robots that blend linear Cartesian motion with articulated rotation to create a new motion type

_____ **35.** A group of gears used to transmit power

_____ **36.** A measure of how fast something is moving

_____ **37.** Revolutions per minute

_____ **38.** A belt that has teeth at set intervals used to transfer power

_____ **39.** Metal items with teeth designed to fit into the links of a chain

_____ **40.** Loss of rotation between the drive and driven pulley

_____ **41.** The diameter of an imaginary circle used to design the gear

_____ **42.** Gears with the same curved tooth of the spiral bevel gear, but the sides are not angled as well

_____ **43.** Another name for helical gears

_____ **44.** An organization that develops, updates, and maintains sets of standards for use by industries of the world

_____ **45.** Motors that move a set portion of the rotation with each application of power

_____ **46.** Rotational force

_____ **47.** A cylinder that has one tooth cut around it

_____ **48.** Systems that consist of a spur gear and a rod or bar that has teeth cut along the length

Matching (1 point each)

Match the terms to the definitions below.

a. Anthropomorphic
b. Ball screw
c. Bevel gears
d. Compound gears
e. Backlash
f. DC brushes
g. Diametral pitch
h. Direct drive
i. Drive pulley
j. Driven pulley
k. Drive system

l. Encoder
m. Flat belt
n. Gear ratio
o. Gear train
p. Green power
q. Harmonic drives
r. Helical gears
s. Helix
t. Hypoid bevel gears
u. Idler gear

_____ **49.** The ratio of the number of teeth per pitch diameter; describes the size of the gear teeth

_____ **50.** The pulley attached to the motor

_____ **51.** The distance from the back of the drive gear tooth to the front of the driven gear tooth

_____ **52.** A flat band of material used to transfer motor power

_____ **53.** Gears similar to spur gears with the teeth set at an angle along the edge instead of parallel

_____ **54.** Two or more gears on the same shaft, often made from one solid piece of material

_____ **55.** A large shaft with a continuous tooth carved along the outer edge with a nut or block that moves up and down the length of the shaft

_____ **56.** A smooth space curve

_____ **57.** Items made of carbon to transfer electricity from the power wires going into the motor to the rotating portions of the motor

_____ **58.** Power derived from a source that is easily renewable that often has little or no environmental impact

_____ **59.** The combination of motors and the moving parts they connect to

_____ **60.** Two or more gears connected together

_____ **61.** Devices that provide feedback about the motor's rotational position

_____ **62.** Specialized gear systems that use an elliptical wave generator to mesh a flex spline with a circular spline that has gear teeth fixed along the interior

_____ **63.** Similar to the spiral bevel gear, but if you draw a line from the shaft set at an angle, it will not meet the shaft of the other gear

_____ **64.** Gears that have their teeth cut along a tapered edge that would make a pointed cone if not flattened on the end

_____ **65.** Systems that have the motor shaft connected directly to the robot for motion

_____ **66.** The ratio we use to determine what happens with the driven gear in reference to the drive gear in terms of torque and speed

_____ **67.** Robots that have motions that look very organic and lifelike

_____ **68.** The pulley attached to the system or output

_____ **69.** Gears added to a system to change the direction of rotation on a dedicated shaft

Short Answer (2 points each)

Write the answer to the following questions in the space provided.

70. Green power source robots are primarily found in which fields of robotics?

_____

_____

_____

71. What is the shape of the work envelope for a cylindrical robot, and what defines the size of this work envelope?

_____

_____

_____

_____

72. What does an ISO certification mean?

_____

_____

_____

_____

_____

73. What is the limiting factor in the payload of a direct drive robot, and what is the current maximum?

_____

_____

_____

_____

_____

74. Define each of the variables in the following equation: $T = F \times R$.

_____

_____

_____

_____

_____

_____

_____

_____

_____

_____

_____

_____

75. What is the level of precision possible with a ball screw?

_____

_____

_____

_____

_____

76. What are the types of wear to watch for with a chain drive system?

_____

_____

_____

_____

_____

_____

77. Describe the work envelope of a SCARA robot.

_____

_____

_____

_____

78. List the three types of belts from the chapter in order from most likely to slip to least likely.

_____

_____

_____

_____

_____

79. What are the two pressure angle choices for gears, and what is the benefit of the newer of the two?

_____

_____

_____

_____

_____

_____

Problems (2 points each)

Answer each of the following questions in the space provided.

80. What is the speed ratio for a system that has a drive pulley running at 1,750 RPM and a driven pulley running at 350 RPM?

_____

_____

_____

_____

_____

_____

81. What HP motor would you need to produce 100 ft. lbs. of torque at 1,750 RPM?

_____

_____

_____

_____

82. What is the velocity of a 48-tooth gear with a diametral pitch of 12 rotating at 100 RPM?

_____

_____

_____

_____

83. What is the required torque for a system that has to move 50 lbs. and has a drive pulley with a diameter of 10 inches?

_____

_____

_____

_____

84. We have a 2.25 HP, 1,750 RPM motor connected to a pulley system. What is the torque, in ft. lbs., that this motor can produce?

_____

_____

_____

_____

85. What would be the radius of the pulley that would move 25 lbs. of material and generate 100 in. lbs. of torque? What would be the diameter of this pulley?

_____

_____

_____

_____

_____

86. What is the pitch diameter of a gear with 45 teeth and a diametral pitch of 15?

_____

_____

_____

_____

_____

_____

87. What is the velocity of a belt system with a drive pulley that has a diameter of 10 in. running at 1,750 RPM?

_____

_____

_____

_____

_____

88. You have a drive pulley that generates 150 in. lbs. of torque and has a pulley diameter of 4 in. What diameter pulley would you need on the driven side to generate 300 in. lbs. of torque? How would this affect the output speed of the system?

_____

_____

_____

_____

_____

_____

89. What HP motor would you need to produce 100 in. lbs. of torque at 1,750 RPM?

_____

_____

_____

_____

_____

90. What is the difference in HP to create 50 ft. lbs. of torque at 1,750 RPM versus 1,250 RPM?

_____

_____

_____

_____

_____

_____

91. What is the center-to-center distance of two gears with gear 1 having a pitch diameter of 6 in. and gear 2 having a pitch diameter of 10 in? Assume that these gears are compatible.

_____

_____

_____

_____

_____

_____

92. What is the gear ratio of a drive gear with 48 teeth and a driven gear with 84 teeth?

_____

_____

_____

_____

_____

93. Given a robot with the specifications below, first determine if the system has enough power to function as designed and then determine the maximum remaining payload capacity for this system. Make sure you show the math to support your answers. For this problem, you do not need to factor in any forces created by motion.

Axis 1: 500 ft. lbs. of torque (due to gearing)
Axis 2: 350 ft. lbs. of torque (due to gearing)
Axis 3: 250 ft. lbs. of torque
Axis 4 and 5: 110 ft. lbs. of torque
Axis 6: 30 ft. lbs. of torque

The tooling weighs 15 lbs., the portion of the robot from axis 5 to 6 weighs 30 lbs., the portion of the robot from axis 4 to 5 weighs 60 lbs., the portion of the robot from axis 3 to 4 weighs 100 lbs., the portion of the robot from axis 2 to 3 weighs 100 lbs., and the portion of the robot from axis 1 to 2 weighs 150 lbs. All the weights include the weight of the robot parts as well as any motors and other equipment installed on the portions of the robot.

_____

_____

_____

_____

_____

94. What would be the diametral pitch of a gear with 50 teeth and a pitch diameter of 4 in.?

_____

_____

_____

_____

_____

_____

95. What is the center-to-center distance of two gears if the first gear has 48 teeth and a diametral pitch of 12 and the second gear has 66 teeth and a diametral pitch of 12?

_____

_____

_____

_____

_____

96. Given a system with a drive pulley with a diameter of 10 in. and a driven pulley with a diameter of 5 in., what is the speed of the driven pulley if we run the drive pulley at 1,000 RPM?

_____

_____

_____

_____

_____

_____

Essay Assignment (12 points)

97. In Chapter 4, we discussed several ways of classifying robots and the reasoning behind each. Of the methods covered, pick the one that you feel is the best and detail the pros and cons of that system. Once you have done this, list the four methods of classification we discussed in order from your favorite to your least favorite and describe why you ordered the classifications in this manner.

RUBRIC	1	2	3	4	Points Earned
CONTENT	The work has one of the following: details about which method was the best, the pros and cons for this method, the four classification methods in order of favorites, and a description of why they are ordered in that manner.	The work is missing only two of the following: details about which method was the best, the pros and cons for this method, the four classification methods in order of favorites, and a description of why they are ordered in that manner.	The work is missing only one of the following: details about which method was the best, the pros and cons for this method, the four classification methods in order of favorites, and a description of why they are ordered in that manner.	The work details which method was the best, lists the pros and cons for this method, has the four classification methods listed in order of favorites, and includes a description of why they are ordered in that manner.	
ORGANIZATION	There is no organization of the material.	There is some organization of the material, but it is still difficult to follow.	The work has a clear structure, but some of the organization interferes with clarity.	The work is clear and concise.	
GRAMMAR	There are four or more spelling, punctuation, or other grammar mistakes.	There are two or three spelling, punctuation, or other grammar mistakes.	There is one spelling, punctuation, or other grammar mistake.	Spelling, punctuation, and grammar are all correct.	

Name: _____   Date: _____

Research Assignment (12 points)

98. In Chapter 4, we examined several ways to classify robots, but by no means did we cover them all. Do some research on your own using what you have learned from this book so far, the materials you have in the classroom, and other reputable sources to learn about another way to classify robots or create your own classification method. Once you have done this, write a report that details how the method classifies robots, the different categories in the classification method with some examples, and how this method compares to the ones listed in the chapter. If you create your own method of classification, make sure you give it plenty of thought and that it is original, not just a spin-off of the methods covered in the chapter. Make sure you document (per your instructor's preference) any resources used.

RUBRIC	1	2	3	4	Points Earned
CONTENT	The report has one of the following: the method, the different categories with examples, a comparison with the methods from the chapter, and any reference material used.	The report is missing two of the following: the method, the different categories with examples, a comparison with the methods from the chapter, and any reference material used.	The report is missing only one of the following: the method, the different categories with examples, a comparison with the methods from the chapter, and any reference material used.	The report has the method, the different categories with examples, a comparison with the methods from the chapter, and any reference material used.	
ORGANIZATION	There is no organization of the material.	There is some organization of the material, but it is still difficult to follow.	The work has a clear structure, but some of the organization interferes with clarity.	The work is clear and concise.	
GRAMMAR	There are four or more spelling, punctuation, or other grammar mistakes.	There are two or three spelling, punctuation, or other grammar mistakes.	There is one spelling, punctuation, or other grammar mistake.	Spelling, punctuation, and grammar are all correct.	

Suggested Lab (value assigned by instructor)

99. Here is your chance to explore the world of robot drive systems with a fun kit known as the Box Robot. You can find the instructions for this lab in the activities manual, and this is a quick and easy build that will add a new robot to your lab arsenal. One of the unique things about this lab is that it uses a worm gear on the motor to transfer power to axle. Many times, we use spur gears and gear trains to get the power from the motor to the outputs, which is why the worm gear is unique. Once you have completed the Box Robot, this system will make a great experimental base for your future explorations of the robot, and I encourage you to find ways to increase the complexity of this robot as your knowledge grows.

Use the provided form to report your experience and earn credit for this lab, unless otherwise directed.

Box Robot Lab Form

Lab Description

In the space below, describe the purpose of the lab and the equipment involved.

_____
_____
_____
_____
_____
_____
_____
_____
_____
_____

Lab Execution

In the space provided, detail the steps you took to perform the lab. Make sure to include any troubleshooting steps performed.

_____
_____
_____
_____
_____
_____
_____
_____
_____
_____
_____

Observations

Record your observations about the systems performance here, including both the expected and unexpected.

_____

_____

_____

_____

_____

_____

_____

_____

_____

_____

Conclusions

What conclusions or statements can you make about the robot based on your observations and any data gathered during the course of the lab?

_____

_____

_____

_____

_____

_____

_____

_____

_____

_____

_____

Name: _____ Date: _____

Score: _____ Text pages 120–143

ACTIVITIES

Multiple Choice (1 point each)

Identify the choice that best completes the statement or answers the question.

_____ 1. To turn a robot into a CNC machine, you need to _____.

    a. add motorized tooling
    b. use proper cutting tools
    c. use proper programming
    d. all of these

_____ 2. To center a part in two directions at once, the recommended gripper is a _____.

    a. two-finger gripper
    b. four-finger gripper
    c. five-finger gripper
    d. vacuum gripper

_____ 3. To allow the robot to use multiple tooling, we can _____.

    a. have the operator physically changed out the tooling
    b. mount multiple tools on the robot
    c. use quick change adaptors and a tool storage system
    d. all of these

_____ 4. Passive RCC systems often have _____.

    a. springs and shear plates
    b. materials that can flex and sensors
    c. alarms tied to the pressure sensor
    d. all of these

_____ 5. When we need a gripper to generate a large amount of force, we generally use _____ as the power supply.

    a. pneumatics
    b. hydraulics
    c. electricity
    d. any of these

_____ **6.** Which type of welding system uses intense beams of light to weld two parts together?

    a. MIG

    b. laser

    c. TIG

    d. arc

_____ **7.** When the process calls for side-to-side centering of the part, the simplest gripper that will work is a _____.

    a. two-finger gripper

    b. four-finger gripper

    c. five-finger gripper

    d. vacuum gripper

_____ **8.** _____ use a coil of copper wire wound around or inside a metallic frame to create an electromagnet.

    a. Vacuum grippers

    b. Magnetic grippers

    c. Parallel grippers

    d. Angular grippers

_____ **9.** _____ have fingers that move in straight lines toward the center or outside of the part to close and grip or open and release the part.

    a. Vacuum grippers

    b. Magnetic grippers

    c. Parallel grippers

    d. Angular grippers

_____ **10.** Grippers with generic fingers _____.

    a. save on cost

    b. allow for in-house machining

    c. must be large enough to have enough material left after machining for structural integrity

    d. all of these

_____ **11.** For odd-shaped parts, the recommended gripper is a _____.

    a. two-finger gripper

    b. three-finger gripper

    c. five-finger gripper

    d. vacuum gripper

_____ **12.** Which of the following would an electromagnet NOT work on?

    a. iron

    b. steel

    c. aluminum

    d. wrought iron

_____ **13.** Vision systems _____.
  a. are replacing RCC devices
  b. are used to offset the system as needed
  c. can sort items by color
  d. all of these

_____ **14.** Which of the follow do we classify as grippers?
  a. three-fingered power open, power closed
  b. electromagnetic systems
  c. vacuum devices
  d. all of these

_____ **15.** _____ create a low-pressure area to pick up parts.
  a. Vacuum grippers
  b. Magnetic grippers
  c. Parallel grippers
  d. Angular grippers

_____ **16.** Which of the following is a downside to quick-change adaptors?
  a. They can make communication and power connections.
  b. They have alignment pins for consistent placement.
  c. They provide positive locking systems.
  d. The initial setup.

_____ **17.** _____ have fingers that hinge or pivot on a point to move the tips outward to release parts or inward to grip parts.
  a. Vacuum grippers
  b. Magnetic grippers
  c. Parallel grippers
  d. Angular grippers

_____ **18.** Which type of welding system uses electricity and feed wire to weld two parts together?
  a. MIG
  b. laser
  c. TIG
  d. arc

Matching (1 point each)

Match the terms to the definitions below.

a. Angular grippers
b. EOAT
c. Ferrous metals
d. Fingers
e. Friction
f. Grippers
g. Jaws
h. Laser welders
i. MIG welders
j. Normal force

k. Odd-shaped parts
l. Parallel gripper
m. Payload
n. RCC
o. Safety factor
p. Tooling base
q. Vacuum
r. Vacuum gripper
s. Welding gun
t. Center of gravity

_____ 19. Systems that use intense beams of light to create the high temperatures needed to join metal

_____ 20. The force resisting the relative motion of two materials sliding against each other

_____ 21. The margin of error we build into a process or system

_____ 22. Another name for tooling fingers

_____ 23. Tools, devices, equipment, and so on at the end of the robotic manipulator

_____ 24. The part of the tooling that attaches to the robot and holds the mechanisms for movement

_____ 25. The part of the gripper that moves to hold parts

_____ 26. Tube-like tools used to direct the force of welding operations

_____ 27. Parts with unique shapes and proportions

_____ 28. Tooling that applies some force to secure parts or objects for maneuvering

_____ 29. The specification of a robotic system that informs the user how much the robot can safely move

_____ 30. A device that works by creating a pressure that is less than atmospheric

_____ 31. Tooling with fingers that move in straight lines toward the center or outside of the part

_____ 32. Metals that contain iron

_____ 33. A machine that uses wire feed through the system and high-current voltage to join two pieces of metal

_____ 34. A simple way for tooling to respond to parts that are not always in the same position

_____ 35. The force that an object pushes back with when acted on by a force

_____ 36. Where we consider the mass to center and all forces in equilibrium

_____ 37. Grippers that have fingers that hinge or pivot on a point to move the tips of the fingers

_____ 38. Pressure that is less than atmospheric

Short Answer (2 points each)

Write the answers to the following questions in the space provided.

39. What is the difference between active and passive RCC units?

_____

_____

_____

_____

_____

_____

_____

40. When we power grippers in only one direction, how is the other motion achieved?

_____

_____

_____

_____

_____

_____

_____

_____

41. How does the bladder-type gripper filled with coffee grounds work?

_____

_____

_____

_____

_____

_____

_____

42. When the robot does not have enough axes to position the tool properly, what are the options to fix this issue?

_____

_____

_____

_____

_____

_____

_____

_____

Problem (2 points each)

Answer each of the following questions in the space provided

43. What is the difference in torque if 5 pounds of weight is positioned 1 inch from a gripper versus 12 inches from the gripper? For this calculation, use the formula $T = W \times d$, where

T = torque

W = weight

d = distance at which weight is applied

    After completing your calculations, what can you conclude about distance and weight as far as it affects gripper force?

_____

_____

_____

_____

_____

_____

_____

_____

44. What is the available payload of a robot that has a maximum payload of 25 kg to which you want to mount two different tools at once, with the first weighing 17 kg and the second weighing 13 kg? Once you have determined the available payload, note any problems you foresee with this system and possible corrective actions for the noted problems.

_____

_____

_____

_____

_____

45. Given that $1 N = 1kg\ m/s^2$ and that N stands for newtons, which is a metric unit of force, and that $W = m \times G$, with W = to weight or the force of an object due to G = gravity multiplied by m = mass, explain how 1 cubic meter of rolled steel has a weight of 76,930 N while 1 cubic meter of lead has a weight of 111,132 N. Remember that the metric unit for gravity is 9.8 m/s². Use mathematical data to back up your explanation.

_____

_____

_____

_____

_____

46. Given the following information, calculate the force required for each jaw to lift a part vertically:

$Fr = W \times G$

$F = \mu N$ or $N = F/\mu$:

- The part weighs 15 pounds.
- The gripper's jaws are parallel.
- The part is gripped 10 inches from the center of gravity.
- The gripping surface is 6 inches long.
- The part is 5 inches wide where it is being gripped.
- The part is being lifted with a maximum acceleration of 1.5 Gs, including normal gravitational force.
- The coefficient of friction between the gripper and the part is 0.90.
- The engineer in charge wants a safety factor of 3 to be included.

_____

_____

_____

_____

_____

_____

47. What would be the required normal force of a gripper that must exert a force of 60 lbs with a coefficient of friction of 0.60? What happens to the normal force for the same system if we increase the friction to 0.80? What does this tell us about friction in gripper systems?

Use the formula $F = \mu N$ or $N = F/\mu$, where

F = friction force

μ = coefficient of friction

N = normal force

_____

_____

_____

_____

_____

_____

48. Given a robot with a maximum payload of 25 kg and tooling that weighs 13 kg, determine the available payload for this system.

_____

_____

_____

_____

_____

_____

49. Given the formula $F = \dfrac{2(m \times G \times d)}{\sqrt{(b^2 + p^2)}}$ and the following information, determine the amount of force that the grippers would need to exert to hold the part and meet the conditions set for the system.

- The part weighs 5 pounds.
- The gripper's jaws are parallel.
- The part is gripped 3 inches from the center of gravity.
- The gripping surface is 1.5 inches long.
- The part is 2 inches wide where it is being gripped.
- The part is being lifted with a maximum acceleration of 2.5 Gs, including normal gravitational force.
- The coefficient of friction between the gripper and the part is 0.85.
- The engineer in charge wants a safety factor of 1.5 to be included.

_____

_____

_____

_____

_____

_____

_____

●
Essay Assignment (12 points)

50. Given what you have learned in the chapter, design a tooling system for a robot that could pick up odd-shaped parts consistently and deal with misalignments. In your system description, make sure you detail what type of tooling you would use, the power source for the tooling, and the various components of your system that deals with misalignments. Along with what you used, make sure you detail why you chose that particular portion for the system.

RUBRIC	1	2	3	4	Points Earned
CONTENT	Only one of the following components is included: outlines a system that would grip the odd-shaped parts consistently and deal with any misalignments, why you chose those parts, and power source for the gripper.	Multiple errors or omissions are seen in the following: outline of a system that would grip the odd-shaped parts consistently and deal with any misalignments, why you chose those parts, and power source for the gripper.	No more than one error or omission is seen in the following: outline of a system that would grip the odd-shaped parts consistently and deal with any misalignments, why you chose those parts, and power source for the gripper.	Outlines a system that would grip the odd-shaped parts consistently and deal with any misalignments, why you chose those parts, and power source for the gripper.	
ORGANIZATION	There is no organization of the material.	There is some organization of the material, but it is still difficult to follow.	The work has a clear structure, but some of the organization interferes with clarity.	The work is clear and concise.	
GRAMMAR	There are four or more spelling, punctuation, or other grammar mistakes.	There are two or three spelling, punctuation, or other grammar mistakes.	There is one spelling, punctuation, or other grammar mistake.	Spelling, punctuation, and grammar are all correct.	

Research Assignment (12 points)

51. Given that there are many other types of robot tooling out there that we did not cover in Chapter 5, use the resources at your disposal such as classroom literature, the library, or reputable Internet resources to track down information and write a brief report on tooling we DID NOT cover in the chapter. Your report should include the tooling's name, how it is used, common power sources, and the basics of its operation. Make sure you list your resources (per your instructor's preferred format) and if possible, include a picture or two of the tooling in your report.

RUBRIC	1	2	3	4	Points Earned
CONTENT	The report has one or two of the following: what the tooling is called, common power sources, the basics of the tooling's operation, how we use it, list of resources used.	The report is missing some of the following: what the tooling is called, common power sources, the basics of the tooling's operation, how we use it, list of resources used.	The report is missing one only of the following: what the tooling is called, common power sources, the basics of the tooling's operation, how we use it, list of resources used.	The report contains what the tooling is called, common power sources, the basics of the tooling's operation, how we use it, list of resources used.	
ORGANIZATION	There is no organization of the material.	There is some organization of material, but it is still difficult to follow.	The work has a clear structure, but some of the organization interferes with clarity.	The work is clear and concise.	
GRAMMAR	There are four or more spelling, punctuation, or other grammar mistakes.	There are two or three spelling, punctuation, or other grammar mistakes.	There is one spelling, punctuation, or other grammar mistake.	Spelling, punctuation, and grammar are all correct.	

Suggested Lab (value assigned by instructor)

52. Since Chapter 5 is all about tooling, this is great time to make some of your own. Find the electromagnet lab in the activities manual and complete it at this time. Once you are finished with this lab, you may want to find ways to use your creation with current classroom labs or robots of your own design. If nothing comes to you now, do not worry, there are more labs down the road where you could use this tooling. You may want to try different cores and different numbers of wraps to see how it changes the performance of the electromagnet or perhaps put the whole assembly in a project box to tidy things up. This is a chance to awaken your inner design engineer and see what you can come up with.

Use the provided lab form to assist with reporting your lab activities, unless directed otherwise by your instructor.

Electromagnet Lab Form
Lab Description

In the space below, describe the purpose of the lab and the equipment involved.

_____

_____

_____

_____

_____

_____

_____

_____

_____

_____

_____

Lab Execution

In the space provided, detail the steps you took to perform the lab. Make sure to include any troubleshooting steps performed.

_____

_____

_____

_____

_____

_____

_____

_____

_____

_____

_____

Observations

Record your observations about the system's performance here, including both the expected and unexpected.

_____

_____

_____

_____

_____

_____

_____

_____

_____

_____

Conclusions

What conclusions or statements can you make about the robot based on your observations and any data gathered during the course of the lab?

_____

_____

_____

_____

_____

_____

_____

_____

_____

_____

CHAPTER 6

Sensors and Vision

Name: _____ Date: _____

Score: _____ Text pages 144–166

ACTIVITIES

Multiple Choice (1 point each)

Identify the choice that best completes the statement or answers the question.

_____ 1. To detect changes in the system, high-end complex tactile sensors may monitor changes in _____ within the sensor due to interaction with their environment.
  a. voltage
  b. resistance
  c. capacitance
  d. all of these

_____ 2. When selecting a limit switch, you need to know _____.
  a. the amperage at the contacts
  b. switch response speed
  c. force exerted by the contact object
  d. all of these

_____ 3. A(n) _____ system sends out control pulses and then receives back a signal that confirms system response.
  a. open-loop
  b. closed-loop
  c. drum-controlled
  d. stepper motor

_____ 4. _____ limit switches have an amperage range of 10 to 20 amps.
  a. Standard and subminiature micro
  b. Standard and micro
  c. Standard and jumbo
  d. Micro and subminiature micro

_____ 5. When should we perform calibration on robotic sensor systems?
  a. never
  b. only during robot installation and verification
  c. only after system damage
  d. after sensor changes, damage to the system, and as recommended by the manufacturer

_____ 6. A(n) _____ system emits a high-frequency sound pulse and measures the amount of time it takes to return to determine the distance between the robot and objects.

    a. GPS

    b. ultrasonic

    c. sound location

    d. vision

_____ 7. Which of the following is not part of the process of setting up a vision system?

    a. taking a calibration image

    b. taking an image of the optimal part position

    c. programming the robot to use the vision system data

    d. none of these

_____ 8. What were some of the problems with early vision systems?

    a. Color was not an option.

    b. Changes in lighting greatly reduced the chance of recognizing objects.

    c. They were prone to failure and thus required human intervention.

    d. All of these.

_____ 9. The _____ is used to count the number of full rotations made as well as ensure proper positioning of the systems should the encoder or motor need adjustment, replacement, or maintenance.

    a. second row of light windows or reflectors

    b. zero pulse

    c. single row of light windows or reflectors

    d. none of these

_____ 10. Optical encoders are prone to failure when _____ gets inside the unit.

    a. oil

    b. metal chips

    c. dust and contaminants

    d. all of these

_____ 11. What is the biggest danger with false alarms using the amperage level sensing method for impact?

    a. someone changing the parameters, leading to impacts of great force

    b. the operator will get in the habit of resetting the alarm and ignore it when the system has an issue

    c. the newer systems are incapable of false alarms

    d. none of these

_____ 12. What are some things, other than an impact, that could trigger an alarm when we use current monitoring for impact?

    a. heavy loads, dirt in the joints, and bad bearings

    b. using tooling lighter than the designated payload

    c. running the system at full speed under normal operating conditions

    d. none of these

_____ 13. Micro limit switches commonly use _____ to activate the change of contacts.

    a. small movements of low force

    b. large movements of large force

    c. small movements of large force

    d. large movements of small force

_____ **14.** A(n) _____ system uses multiple microphones and special programming to pinpoint the location of sounds.

  a. GPS

  b. ultrasonic

  c. sound location

  d. vision

_____ **15.** To reduce the force the robot has to stop, the energy of the impact, and the stress on the robot's internal system, we _____.

  a. E-stop the robot when alarm conditions exist

  b. remove all power to the system

  c. disengage the axes

  d. reverse the direction of all the motors as soon as an impact is detected

_____ **16.** Complex tactile sensors can detect _____.

  a. applied force

  b. shape of the part

  c. temperature

  d. any of these

_____ **17.** If we choose the wrong color of light to illuminate parts for a vision system, _____.

  a. it can make certain colors nearly invisible

  b. it may complicate the process

  c. it may hide key features of the part

  d. all of these

_____ **18.** Open-loop control systems are popular in _____.

  a. servo motor-driven robots

  b. most industrial systems

  c. electric only robots

  d. early pneumatic and stepper motor systems

_____ **19.** The _____ for a tactile array is the signal generated by the element under no contact conditions.

  a. calibration signal

  b. mastering signal

  c. base signal

  d. none of these

_____ **20.** _____ limit switches have an amperage range of 1 to 7 amps for the contacts.

  a. Standard

  b. Micro

  c. Jumbo

  d. Subminiature micro

_____ 21. A(n) _____ system uses the time it takes to receive signals from three or four separate satellites in orbit around the Earth to determine position.

    a. GPS

    b. ultrasonic

    c. sound location

    d. vision

_____ 22. Initially, _____ was all about keeping the robot and other equipment safe.

    a. mastering

    b. impact detection

    c. calibration

    d. complex tactile sensing

_____ 23. Limit switches come in all the following sizes except _____.

    a. standard

    b. micro

    c. jumbo

    d. subminiature micro

_____ 24. If we use a red light to illuminate parts for a vision system, all items _____ in color will be nearly invisible.

    a. red

    b. blue

    c. green

    d. white

_____ 25. A(n) _____ system uses one or more cameras and special software to turn the images taken by the camera into useful data for the robot.

    a. GPS

    b. ultrasonic

    c. sound location

    d. vision

Matching (1 point each)

Match the terms to the definitions below.

a.	Absolute optical encoder	**l.**	Limit switch
b.	Binary address	**m.**	Motor encoder
c.	Calibration	**n.**	Noise
d.	Capacitive proximity switch	**o.**	Open loop
e.	Closed loop	**p.**	Photoelectric proximity switch
f.	Eddy currents	**q.**	Proximity switch
g.	GPS	**r.**	Solid-state device
h.	Hall effect sensor	**s.**	Tactile
i.	Impact	**t.**	Ultrasonic sensors
j.	Incremental optical encoder	**u.**	Vision system
k.	Inductive proximity switch		

_____ **26.** A unique set of 1s and 0s that the controller can understand and use

_____ **27.** Devices made up of a solid piece of material that manipulates the flow of electrons; without any moving parts

_____ **28.** Systems that send out the control pulse to initiate movement and then receive back a signal that confirms motor rotation, direction, and distance traveled

_____ **29.** Switches that send out a specific wavelength of light and use a receiver to detect that specific wavelength of light

_____ **30.** Devices activated by contact with an object and change the state of their contacts when the object exerts a certain amount of force

_____ **31.** A flow of electrons created by a magnetic field moving across a ferrous metal item, which in turn causes the metal item to generate its own magnetic field of an opposite polarity

_____ **32.** Systems that use cameras and software to process images and provide that information to the robot

_____ **33.** A sensor that uses a magnetic field to cause voltage flow in a semiconductor; used to track motor rotation

_____ **34.** Solid-state devices that use light, magnetic fields, or electrostatic fields to detect various items without the need for physical contact

_____ **35.** Devices that directly monitor the rotation of a motor shaft and turn that information into a meaningful signal

_____ **36.** Mathematical difficulties in properly calculating the torque required at the start of motor motion for robot movement

_____ **37.** A disk that has either holes for light to pass through or special reflectors to return light, an emitter, a receiver, and some solid state devices for signal interpretation and transmission

_____ **38.** The ability to sense pressure and impact

_____ **39.** A system that determines geographical position based on the time it takes to receive a signal from three or four separate satellites in orbit around the Earth

_____ **40.** A switch that uses an oscillating magnetic field to detect ferrous metal items

_____ **41.** A specified process that ensures a precision system performs properly and provides for any adjustments needed

_____ **42.** Systems that work on the assumption that the control pulse activates the motion system and the robot performs as expected

_____ **43.** Robot contact with an object in the intended movement path

_____ **44.** Switches that generate an electrostatic field and work on the same principle as capacitors

_____ **45.** Sensors that emit and receive sound to sense items

_____ **46.** Encoders with enough emitters and receivers to give each position of the encoder its own unique binary address

Short Answer (2 points each)

Write the answers to the following questions in the space provided.

47. What is the difference between a light level photoelectric prox switch and a color detection photoelectric switch?

_____

_____

_____

_____

_____

_____

_____

_____

_____

48. How do we use amperage levels to sense impact?

_____

_____

_____

_____

_____

_____

_____

_____

_____

_____

_____

_____

_____

_____

49. What are the common voltage ranges for DC, AC, and AC/DC combo prox switches?

_____

_____

_____

_____

_____

_____

50. What is the Hall effect?

_____

_____

_____

_____

_____

_____

_____

_____

51. How do robots use ultrasonic sensors to find leaks in pressure vessels?

_____

_____

_____

_____

_____

_____

_____

_____

52. What are the downsides to a robot going into E-stop and locking all the axes in place when an impact is detected?

_____

_____

_____

_____

_____

_____

53. What is the difference in information received from a simple grip detection tactile sensor and a complex tactile sensor?

_____

_____

_____

_____

_____

_____

54. If you have a limit switch in the field that has the actuator in the wrong position, how would you correct the situation?

_____

_____

_____

_____

_____

Essay Assignment (12 points)

55. Using the information learned in Chapter 6, identify the various sensors used by your classroom's robotic systems. If you have multiple systems in the classroom, make sure to do this evaluation for all the systems you will be working with during the course. Make a list of the sensors you identify, detailing the type of sensors, the number in use, how they are used, and the type of information they send to the robot. Save this information for later, as it could prove very useful during programming.

RUBRIC	1	2	3	4	Points Earned
CONTENT	The work contains a partial list of the robot's sensors and details some of the following: type, number, and type of information the sensor provides (for each robot listed).	The work is missing some of the following: list of the robot's sensors and type, number, and type of information the sensor provides (for each robot listed).	The work contains a nearly complete list of the robot's sensors and type, number, and type of information the sensor provides (for each robot listed).	The work contains a complete list of the robot's sensors and type, number, and type of information the sensor provides (for each robot listed).	
ORGANIZATION	There is no organization of the material.	There is some organization of material, but it is still difficult to follow.	The work has a clear structure, but some of the organization interferes with clarity.	The work is clear and concise.	
GRAMMAR	There are four or more spelling, punctuation, or other grammar mistakes.	There are two or three spelling, punctuation, or other grammar mistakes.	There is one spelling, punctuation, or other grammar mistake.	Spelling, punctuation, and grammar are all correct.	

Research Assignment (12 points)

56. In Chapter 6, we mentioned resolvers and synchros, a type of position tracking device like the encoder, but we did not go any deeper than that. This is your chance to dig into these systems. Using materials available from your classroom, library, and reputable Internet resources, pick one of the two systems and write a report detailing which system you chose, how it works, the pros and cons of the system, and how it compares in operation to the encoders discussed in the chapter. Make sure you list where (per your instructor's desired format) you gathered your information from.

RUBRIC	1	2	3	4	Points Earned
CONTENT	The report has one of the following: the system chosen, how it works, pros and cons of the system, how it compares to the encoders from the chapter, and a list of references.	The report is missing two of the following: the system chosen, how it works, pros and cons of the system, how it compares to the encoders from the chapter, and a list of references.	The report is missing only one of the following: the system chosen, how it works, pros and cons of the system, how it compares to the encoders from the chapter, and a list of references.	The report details the system chosen, how it works, pros and cons of the system, how it compares to the encoders from the chapter, and a list of references.	
ORGANIZATION	There is no organization of the material.	There is some organization of material, but it is still difficult to follow.	The work has a clear structure, but some of the organization interferes with clarity.	The work is clear and concise.	
GRAMMAR	There are four or more spelling, punctuation, or other grammar mistakes.	There are two or three spelling, punctuation, or other grammar mistakes.	There is one spelling, punctuation, or other grammar mistake.	Spelling, punctuation, and grammar are all correct.	

Suggested Lab (value assigned by instructor)

57. Since Chapter 6 is all about the robot knowing what is going on in the world around it, this is a great time to use sensors in a robot build. The Obstacle Avoiding Robot Lab in the activities manual will walk you through a lab where you build a simple robot that can respond to its environment. The sensors used in this lab are two micro limit switches, and this is a great way to start your sensor exploration journey. Pay close attention to how you wire the switches to reverse the power flow to the motors and thus the motion of the robot, as you can use this trick for other robots. This is another great robot base for future experiments either in the classroom or on your own.

Use the provided lab form to assist with reporting your lab activities, unless directed otherwise by your instructor.

Obstacle Avoiding Robot Lab Form
Lab Description

In the space below, describe the purpose of the lab and the equipment involved.

_____
_____
_____
_____
_____
_____
_____
_____
_____
_____
_____

Lab Execution

In the space provided, detail the steps you took to perform the lab. Make sure to include any troubleshooting steps performed.

_____
_____
_____
_____
_____
_____
_____
_____
_____
_____
_____

Name: _____ Date: _____

Observations

Record your observations about the system's performance here, including both the expected and unexpected.

_____

_____

_____

_____

_____

_____

_____

_____

_____

_____

Conclusions

What conclusions or statements can you make about the robot based on your observations and any data gathered during the course of the lab?

_____

_____

_____

_____

_____

_____

_____

_____

_____

Name: _____ Date: _____

Score: _____ Text pages 167–185

ACTIVITIES

Multiple Choice (1 point each)

Identify the choice that best completes the statement or answers the question.

_____ 1. What are the components of the core robotic system?
- **a.** the controller and external positioner
- **b.** the robot, controller, and power supply
- **c.** the teach pendant, robot, and tooling
- **d.** all of these

_____ 2. Which of the following is NOT a peripheral system associated with tooling?
- **a.** slag cleaning systems
- **b.** force sensors
- **c.** mobile bases
- **d.** polishing systems

_____ 3. One of the main benefits of building a base model robot that works with several peripheral systems is _____.
- **a.** lower total cost of the system
- **b.** the need to create many different types of robots
- **c.** the manufacturer does not have to invest in research and development
- **d.** none of these

_____ 4. Real-time data monitoring capability added to a network _____.
- **a.** allows engineers and maintenance to monitor equipment from networked terminals
- **b.** is rarely used these days
- **c.** is primarily added to help the machine or robot operator
- **d.** all of these

_____ 5. Which of the following is NOT a characteristic of peripheral systems?
- **a.** They feed data to the robot.
- **b.** They perform tasks based on information from the robot.
- **c.** You could remove them from the robot and still have a functioning system.
- **d.** None of these.

_____ **6.** Which of the following are ways networks keep data from being lost due to multiple transmissions all happening at once?

    a. checking for transmissions in progress before sending

    b. passing a digital token

    c. using master and slave setups

    d. all of these

_____ **7.** Connecting a network of highly automated systems to the Internet _____.

    a. allows engineers and maintenance to monitor equipment from home

    b. allows for automated startup and change of production runs

    c. enables the system to send texts or emails under preset conditions

    d. all of these

_____ **8.** A worst-case scenario for communication would be _____.

    a. total signal loss

    b. slow data communications

    c. when some signals make it through correctly and some are distorted or totally lost

    d. having the signal sent to all the control devices on the network

_____ **9.** Mobile robot bases can _____.

    a. keep up with moving parts on chains and conveyors

    b. feed multiple machines in an area

    c. move parts to and from warehouses

    d. all of these

_____ **10.** Which of the following is NOT a benefit of the work cell?

    a. a reduction in the number of times a part must be handled

    b. an increase in overall part cost

    c. a reduction in travel distance between parts

    d. cost savings to the customer

_____ **11.** Who is responsible for the safety protocols of an industrial robot?

    a. RIA

    b. OSHA

    c. the manufacturer

    d. the buyer

_____ **12.** The oldest method of getting information from sensors and other machines to the robot is _____.

    a. direct-wired connection

    b. wireless network

    c. USB

    d. wireless modem

_____ **13.** Which of the following is true about USB communication?

    a. USB communication predates the RS232 standard

    b. USB communication is fast, with fewer problems with pin swapping

    c. USB connections can share large amounts of power over the wire

    d. all of these

Matching (1 point each)

Match each type of network to the characteristics below.

    a. Hardwired network   **b.** Wireless network

_____ **14.** Requires a physical connection between the devices in the network

_____ **15.** May use USB connections

_____ **16.** May use Bluetooth technology

_____ **17.** Communicates without the need for physical wiring between the devices

_____ **18.** May require only a module or card to interface with other systems

_____ **19.** May use RS232 cables

Matching (1 point each)

Match the terms to the definitions below.

    a. Analog signals   **e.** Peripheral systems

    b. Digital signals   **f.** Programmable Logic Controller (PLC)

    c. Lean manufacturing   **g.** RS232

    d. Networking   **h.** Work cell

_____ **20.** Signals that have exactly two states, 1 or 0

_____ **21.** Systems or equipment that perform tasks related to or involving the robot but are not part of the robot

_____ **22.** An initiative that is all about cutting production costs

_____ **23.** A logical grouping of machines that perform various operations on parts in a logical order as a part of the production process

_____ **24.** A communication standard in which certain wires are designated to transmit specific signals

_____ **25.** Specialized computers with logical programs to direct equipment action based on inputs and other information filtered by the program

_____ **26.** Signals with a range of voltage or amperage that correlate to a set scale

_____ **27.** Two or more systems sharing information over some form of connection

Matching (1 point each)

Match each signal type to the devices and uses below.

   a. Analog   **b.** Digital

_____ **28.** You need to determine the flow rate of a powder

_____ **29.** You need to monitor the temperature in a tank

_____ **30.** The type of signal often used for part presence is _____

_____ **31.** For yes- or no-type questions

_____ **32.** A level sensor generates _____ signals

_____ **33.** Most proximity switches generate _____ signals

_____ **34.** To take a range measurement between the robot and an object

_____ **35.** The type of signal generated by a limit switch

Short Answer (2 points each)

Write the answers to the following questions in the space provided.

36. What are some of the tasks that positioning devices perform?

_____
_____
_____
_____
_____
_____
_____
_____

37. How do self-charging robots find their way to the charging station?

_____
_____
_____
_____
_____
_____
_____
_____

38. How do we determine if a mobile base is a part of the robot or a peripheral system?

_____
_____
_____
_____
_____
_____
_____
_____

39. What are the two broad ways we network systems together?

_____
_____
_____
_____
_____
_____

40. What is the connection type that is replacing the RS232 communication standard and why?

_____
_____
_____
_____
_____
_____
_____

Problem (2 points)

Answer the following question in the space provided.

41. Below is a picture of a work cell that has been labeled with arrows and numbers. Each arrow has the tip touching a part of the work cell. It is your task to list what each arrow is pointing to and then detail whether it is a part of the core robot or a peripheral.

Essay Assignment (12 points)

42. Now that you understand what a peripheral system is, take the time to identify those used by the robots you are working with. Make sure you detail what the peripheral is, the job it performs, and how the robot communicates with said device. Make sure to keep this information handy for programming projects down the road.

RUBRIC	1	2	3	4	Points Earned
CONTENT	The essay has one of the following: list of all peripheral equipment used by the robot in the classroom, the job it performs, and how the robot communicates with it.	The essay is missing two of the following: list of all peripheral equipment used by the robot in the classroom, the job it performs, and how the robot communicates with it.	The essay is missing only one of the following: list of all peripheral equipment used by the robot in the classroom, the job it performs, and how the robot communicates with it.	The essay lists all peripheral equipment used by the robot in the classroom, the job it performs, and how the robot communicates with it.	
ORGANIZATION	There is no organization of the material.	There is some organization of the material, but it is still difficult to follow.	The work has a clear structure, but some of the organization interferes with clarity.	The work is clear and concise.	
GRAMMAR	There are four or more spelling, punctuation, or other grammar mistakes.	There are two or three spelling, punctuation, or other grammar mistakes.	There is one spelling, punctuation, or other grammar mistake.	Spelling, punctuation, and grammar are all correct.	

Research Assignment (12 points)

43. In Chapter 7, we talked about peripheral systems but by no means covered them all, and most we only covered briefly. Take some time to use resources such as literature from your classroom, the library, or the Internet and find a specific peripheral system. Write a report about the peripheral you chose, detailing what it is called, how it works, how it benefits the robot, and how it communicates with the robot (if applicable). If possible, provide a picture and some examples of how the system is used. Make sure to cite your references per your instructor's desired format/instructions.

RUBRIC	1	2	3	4	Points Earned
CONTENT	The report has one of the following: the name of the system, how it works, how it benefits the robot, how it communicates with the robot, and references.	The report is missing two of the following: the name of the system, how it works, how it benefits the robot, how it communicates with the robot, and references.	The report is missing one of the following: the name of the system, how it works, how it benefits the robot, how it communicates with the robot, and references.	The report has the name of the system, how it works, how it benefits the robot, how it communicates with the robot, and references.	
ORGANIZATION	There is no organization of the material.	There is some organization of the material, but it is still difficult to follow.	The work has a clear structure, but some of the organization interferes with clarity.	The work is clear and concise.	
GRAMMAR	There are four or more spelling, punctuation, or other grammar mistakes.	There are two or three spelling, punctuation, or other grammar mistakes.	There is one spelling, punctuation, or other grammar mistake.	Spelling, punctuation, and grammar are all correct.	

Suggested Lab (value assigned by instructor)

44. This is a good time to work with the lab equipment you have in the classroom and dig deeper into how it all works. Check with your instructor and see if there is a lab where you can interact with or create external systems for the robot(s) you are working with. Ideally, the lab will consist of one or more peripheral systems that you will connect to the main robot in some fashion and then use to perform activities in the lab. If you are working with VEX or LEGO NXT systems, you can look through the provided build instructions or search the Internet to find instructions that include external systems. You could even create your own external system based on what you have learned up to this point in the course. Given that many robots used in industry and other places work with external systems, the more you can learn about this topic, the better prepared you will be.

Use the provided lab form to assist with reporting your lab activities, unless directed otherwise by your instructor.

Peripheral Equipment Lab Form

Lab Description

In the space below, describe the purpose of the lab and the equipment involved.

_____

_____

_____

_____

_____

_____

_____

_____

_____

_____

_____

_____

Lab Execution

In the space provided, detail the steps you took to perform the lab. Make sure to include any troubleshooting steps performed.

_____

_____

_____

_____

_____

_____

_____

_____

_____

_____

_____

_____

Observations

Record your observations about the system's performance here, including both the expected and unexpected.

_____

_____

_____

_____

_____

_____

_____

_____

_____

_____

Conclusions

What conclusions or statements can you make about the robot based on your observations and any data gathered during the course of the lab?

_____

_____

_____

_____

_____

_____

_____

_____

_____

Name: _____ Date: _____

Score: _____ Text pages 186–207

ACTIVITIES

Multiple Choice (1 point each)

Identify the choice that best completes the statement or answers the question.

_____ 1. If we do not check the fixture after a crash that causes positional issues, which of the following could occur?

    a. Nothing; the fixture position is irrelevant.

    b. Changing the fixture could affect the program that was running during the crash.

    c. Changing the fixture could affect the positioning of all the programs.

    d. None of these.

_____ 2. _____ between motor encoders and the robot controller can cause the robot to move a large distance at full or near full speed unexpectedly.

    a. E-stops

    b. Intermittent communication

    c. Digital signals

    d. Normal communication

_____ 3. Which of the following would NOT fall under the category of robot damage?

    a. dents in the robot

    b. the robot is E-stopped

    c. broken tooling

    d. communication cables with a cut in the insulation

_____ 4. Which of the following is NOT occurring during the boot sequence?

    a. All available data is checked.

    b. The necessary logical sorting to determine the robot state occurs.

    c. The program begins running.

    d. All computational systems get up and running.

_____ 5. If a robot system is tied to five different E-stops, how many times do you need to go through the E-stop checking procedure?

    a. two or three times

    b. at least once

    c. three or four times

    d. at least five times

_____ 6. When using brake release switches to move the robot out of a crash condition, _____.

     a. the robot may suddenly collapse under its own weight

     b. there is nothing left to hold the released axis or axes in place

     c. you must be very careful to select the correct axis and be aware of your body position in relation to the robot

     d. all of these

_____ 7. When do we remove or tighten the communication connections on a robot?

     a. any time

     b. only when the power is on

     c. only when the power is off

     d. anytime we suspect problems

_____ 8. What could happen if you try to hide parts damaged in a robot crash?

     a. There may not be enough parts to fill the order.

     b. You may get a verbal or written reprimand.

     c. You could be fired.

     d. All of these.

_____ 9. Which of the following is NOT a precaution to keep in mind when starting the running program of a robot?

     a. Make sure the robot is in the home position.

     b. Watch at least one full cycle.

     c. Make sure the system does not have an alarm condition.

     d. None of these.

_____ 10. Critical alarms _____.

     a. prevent the machine from running until corrected

     b. do not stop the robot

     c. are very common

     d. all of these

_____ 11. The right-hand rule is designed to work with the _____ frame of the robot.

     a. Joint

     b. World

     c. User

     d. Tool

_____ 12. A common way we define the function of a robot is by _____.

     a. where we use it

     b. how we use it

     c. the task it performs

     d. all of these

_____ 13. The two main things we do in manual mode are _____

     a. move the robot and check program operation

     b. move the robot and run programs automatically

     c. run at full speed and check the program

     d. move the robot at maximum speed and run programs automatically

_____ 14. Which of the following conditions would stop a robotic system running a program in automatic?

    a. The program instructs the robot to stop or reaches the end of the instructions.

    b. The user hits stop, pause, or E-stop.

    c. An alarm condition occurs.

    d. All of these.

_____ 15. Which of the following is considered the worst-case scenario for moving the robot out of a crash position?

    a. using manual mode in the Joint frame

    b. using brake release buttons

    c. using manual mode in the World frame

    d. physically removing motors or turning gears by hand with the power off

_____ 16. If the crash has enough force to move the axes of the robot, which of the following is a possible result?

    a. All the programs will have positional issues.

    b. Only the program running during the crash will have positional issues.

    c. The robot will auto correct for this if you home the system.

    d. None of these.

_____ 17. Which of the following is a hydraulic system–specific check and NOT general fluid power check?

    a. ensuring you have the proper pressure

    b. making sure there is proper fluid level and any cooling systems are working

    c. checking for any leaks in the system

    d. checking and changing filters as needed

_____ 18. When moving the axes as part of our powered up checks, which of the following is NOT true?

    a. This should be done in the automatic mode.

    b. One of the things you are looking for is jerky motion.

    c. You should listen for odd sounds during the movements.

    d. You are looking for any noticeable changes in motion.

_____ 19. When cleaning the robot, _____.

    a. it is a good idea to power the robot down

    b. the goal is to remove buildups of dust or grime

    c. avoid using compressed air

    d. all of these

_____ 20. Designating the running program _____.

    a. is the same for every system

    b. is above the task level of most operators

    c. is as varied as the manufacturers of robotic systems

    d. none of these

_____ 21. Once you are done with manual movements and checks, what should you do to ensure proper robot operation?

    a. make sure the speed is set to 5 percent

    b. make sure the robot is back at home or in a safe position

    c. make sure the robot is in manual

    d. all of these

Matching (1 point each)

Match the crash recovery step numbers to the instructions below.

a. First step   **d.** Fourth step

b. Second step   **e.** Fifth step

c. Third step

_____ **22.** Determine how to prevent another crash.

_____ **23.** Check the alignment of all equipment involved in the crash.

_____ **24.** Get the robot clear of the crash or impact area.

_____ **25.** Determine what to do with the parts involved in the crash.

_____ **26.** Determine why the robot crashed.

Matching (1 point each)

Match the terms to the definitions below.

a. Automatic mode   **h.** Manual mode

b. Continuous mode   **i.** Mentoring

c. Coordinate systems   **j.** Right-hand rule

d. Crash   **k.** Slag

e. Critical alarms   **l.** Step mode

f. Frames   **m.** User frame

g. Joint frame   **n.** World frame

_____ **27.** Error conditions that prevent the machine from running until corrected

_____ **28.** A condition in which the robot moves from line to line of code, requiring some action from you to progress from the current line to the next

_____ **29.** Knowledge shared through teaching, coaching, and helping with experiences as they happen

_____ **30.** A way of referencing movements and points in the work envelope that controls how the robot moves

_____ **31.** Molten metal splatter that has hardened in the welding gun tip

_____ **32.** A condition in which you maintain control of the robot operations and have the ability to stop action as quickly as you can react

_____ **33.** A specially defined Cartesian-based system in which the user defines the zero point and how the positive directions of the axes lay

_____ **34.** A condition in which the program runs as it would normally, provided you maintain the proper inputs such as the dead man switch and manual button

_____ **35.** A condition in which the robot runs the program without your continued input, based on the operating parameters of the program

_____ **36.** A way to determine how the robot will move while using a Cartesian-based frame motion by using the thumb, forefinger, and middle finger.

_____ **37.** Another term for "frames"

_____ **38.** A Cartesian system based on a point in the work envelop where the robot base attaches

_____ **39.** Unexpected contact of the robot with something in the work envelope

_____ **40.** A reference system that moves one axis at a time with the positive and negative directions determined by the setup of each axis's zero point

Matching (1 point each)

Match each action below to the proper timing.

a. Before startup   b. After startup

_____ **41.** Checking E-stops for proper operation
_____ **42.** Checking sensors for proper LED indication
_____ **43.** Checking communication cables for tight connection
_____ **44.** Checking the robot for damage
_____ **45.** Checking the safety equipment
_____ **46.** Checking the ground on a welding robot
_____ **47.** Checking for alarm conditions
_____ **48.** Cleaning the robot

Short Answer (2 points each)

Write the answers to the following questions in the space provided.

49. What are some types of damage that can occur during a crash?

_____
_____
_____
_____
_____
_____
_____
_____

50. What do we want to avoid during the boot process, and what could happen if we rush things?

_____
_____
_____
_____
_____
_____
_____
_____
_____

51. What is the process of checking E-stops?

_____

_____

_____

_____

_____

_____

_____

_____

52. What are some of the pre–power up checks for nonindustrial robots?

_____

_____

_____

_____

_____

_____

_____

_____

_____

_____

_____

_____

53. What are some of the conditions that necessitate running the robot in manual before trying automatic operation?

_____

_____

_____

_____

_____

_____

_____

_____

_____

54. What is the point behind powered robot checks?

_____

_____

_____

_____

_____

_____

_____

55. If you move the robot to access cables, what must you do before you begin working on the system, and what could happen if you forget this step?

_____

_____

_____

_____

_____

_____

_____

56. What are some of the parts of the robot that make use of communication connections?

_____

_____

_____

_____

_____

_____

_____

57. What do you do if you are unsure why a robot crashed?

_____

_____

_____

_____

_____

_____

_____

58. What is the danger of resetting alarms such as battery life expired or tooling expired and not taking care of the problem?

_____

_____

_____

_____

_____

_____

_____

Problems (2 points each)

Answer each of the following questions in the space provided.

59. Label the diagram below, associating each finger to an axis of the robot and whether it is the positive or negative direction using the right-hand rule.

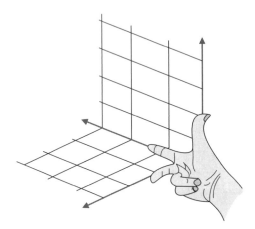

Essay Assignment (12 points)

60. Now that you have an idea of the various checks necessary to ensure proper operation, it is your turn to determine the specific checks required for the robot(s) you are working with in the classroom. Create a list of tasks to perform before powering the system up and another of tasks to perform after startup. Do not be afraid to add something to the list that is not done currently, if you feel it is justified. Once you have created the two lists, write a brief description of how you accomplish these tasks and why each task is important.

RUBRIC	1	2	3	4	Points Earned
CONTENT	The work has one of the following: a list of prepower checks, a list of checks after startup, a description of how to perform each check, and a statement of why each check is important.	The work is missing two of the following: a list of prepower checks, a list of checks after startup, a description of how to perform each check, and a statement of why each check is important.	The work is missing only one of the following: a list of prepower checks, a list of checks after startup, a description of how to perform each check, and a statement of why each check is important.	The work contains a list of prepower checks, a list of checks after startup, a description of how to perform each check, and a statement of why each check is important.	
ORGANIZATION	There is no organization of the material.	There is some organization of the material, but it is still difficult to follow.	The work has a clear structure, but some of the organization interferes with clarity.	The work is clear and concise.	
GRAMMAR	There are four or more spelling, punctuation, or other grammar mistakes.	There are two or three spelling, punctuation, or other grammar mistakes.	There is one spelling, punctuation, or other grammar mistake.	Spelling, punctuation, and grammar are all correct.	

Name: _____ Date: _____

Research Assignment (12 points)

61. In Chapter 8, we talked about the various checks that may be required both before and after startup, but we looked at this only in broad terms. Here is your chance to dig into a specific system and discover the specific checks of that system. Using resources such as the library, Internet, or material in your classroom, select a robotic system used in industry (besides any you are using for lab in the classroom) and write a report about the recommended checks for that system. Make sure you detail which system you chose and whether the checks are before or after startup, and include any special precautions associated with that check if applicable. After you have done this, briefly compare this system to either a system you are using in the classroom or one of the general examples mentioned in the book, noting the similarities and differences. Make sure you add a section of references for any research materials used.

RUBRIC	1	2	3	4	Points Earned
CONTENT	The report has one of the following: the chosen robot system, the prestartup checks, the poststartup checks, a comparison with similarities and differences between the chosen system and a lab- or book-described robotic system, and a research reference section.	The report is missing two of the following: the chosen robot system, the prestartup checks, the poststartup checks, a comparison with similarities and differences between the chosen system and a lab- or book-described robotic system, and a research reference section.	The report is missing only one of the following: the chosen robot system, the prestartup checks, the poststartup checks, a comparison with similarities and differences between the chosen system and a lab- or book-described robotic system, and a research reference section.	The report lists the chosen robot system, the prestartup checks, the poststartup checks, a comparison with similarities and differences between the chosen system and a lab- or book-described robotic system, and a research reference section.	
ORGANIZATION	There is no organization of the material	There is some organization of the material, but it is still difficult to follow	The work has a clear structure, but some of the organization interferes with clarity.	The work is clear and concise.	
GRAMMAR	There are four or more spelling, punctuation, or other grammar mistakes.	There are two or three spelling, punctuation, or other grammar mistakes.	There is one spelling, punctuation, or other grammar mistake.	Spelling, punctuation, and grammar are all correct.	

Suggested Lab (value assigned by instructor)

62. Here is your chance to practice what you have learned. Use either the list of tasks you created for 60, the Essay question, or one provided by your instructor to perform both pre–power up checks and startup checks to ensure proper system operation. Pay careful attention during these checks for anything that indicates a problem and correct any problems found as best you can. If you run into a problem you cannot correct or are unsure how to fix, ask your instructor for direction. At the end of this lab, the robot should be ready to run and if available, load a program and run the system to verify operation. These tasks are done all over the country at the beginning of each workday, and this lab will give you an idea of what it takes to get the robot ready to go.

Use the provided lab form to assist with reporting your lab activities, unless directed otherwise by your instructor.

Preparing the Robot to Run
Lab Form
Lab Description

In the space below, describe the purpose of the lab and the equipment involved.

_____

_____

_____

_____

_____

_____

_____

_____

_____

_____

Lab Execution

In the space provided, detail the steps you took to perform the lab. Make sure to include any troubleshooting steps performed.

_____

_____

_____

_____

_____

_____

_____

_____

_____

_____

Observations

Record your observations about the system's performance here, including both the expected and unexpected.

_____

_____

_____

_____

_____

_____

_____

_____

_____

_____

Conclusions

What conclusions or statements can you make about the robot based on your observations and any data gathered during the course of the lab?

_____

_____

_____

_____

_____

_____

_____

_____

_____

_____

CHAPTER 9 — Programming and File Management

Name: _____ Date: _____

Score: _____ Text pages 208–237

ACTIVITIES

Multiple Choice (1 point each)

Identify the choice that best completes the statement or answers the question.

_____ **1.** Subroutines _____.

    a. reduce the lines of code in a program

    b. make it easier to write programs

    c. include such things as opening or closing grippers, tool changes, and alarm response protocols

    d. all of these

_____ **2.** What do you normally have to do with the teach pendant to run a robot in manual mode?

    a. press the E-stop

    b. place the robot in automatic and hit the start button

    c. hold the dead man switch while pressing and possibly holding a key or combination of keys on the key pad

    d. release the dead man switch and hit reset

_____ **3.** Robots move slower in manual mode _____.

    a. for safety reasons

    b. because of the greater processing requirements on the controller

    c. only on level 1 programmed systems

    d. only when you create a new program

_____ **4.** _____ motion is arcs or full circles that require at least three points to execute.

    a. Joint

    b. Linear

    c. Circular

    d. Weave

_____ 5. _____ motion is where the robot moves in a straight line between two points.
- **a.** Joint
- **b.** Linear
- **c.** Circular
- **d.** Weave

_____ 6. Which of the following do we use to back up robot programs to prevent loss?
- **a.** SD cards
- **b.** hard drives
- **c.** flash drives
- **d.** any of these

_____ 7. Before deleting a robot program, it is wise to _____.
- **a.** remove the storage device so that no trace is left
- **b.** start a new program with the same name
- **c.** save a properly labeled copy somewhere external to the robot
- **d.** save a copy in the robot's short-term memory

_____ 8. An example of _____ programming would be changing the pegs on a rotating drum to modify robot operation.
- **a.** level 1
- **b.** level 2
- **c.** level 3
- **d.** level 4

_____ 9. In _____ programming, it is common practice for the programmer to write the basic program offline, with each position having a label but no coordinate data.
- **a.** level 1
- **b.** level 2
- **c.** level 3
- **d.** level 4

_____ 10. Which of the following is NOT true of scheduled automatic robot data backups?
- **a.** They ensure data is not lost.
- **b.** They allow you to recall previous versions of programs.
- **c.** They are rarely used in industry.
- **d.** Most systems allow the user to determine when old saves are overwritten.

_____ 11. _____ motion is the addition of angular side-to-side movement while performing other base motion types.
- **a.** Joint
- **b.** Linear
- **c.** Circular
- **d.** Weave

_____ 12. To reprogram a robot that has a level 1 programming language, we _____.

     a. change the program on the main computer and download it into the robot

     b. physically change something in the system

     c. use the teach pendant to make changes

     d. put the system in teach mode and then physically show the robot what we want it to do

_____ 13. _____ motion is point-to-point motion with no correlation between the axes.

     a. Joint

     b. Linear

     c. Circular

     d. Weave

_____ 14. Which of the following is NOT something that occurs during normal operation and requires the programmer to modify or fine-tune the robot program?

     a. variances in batches of parts

     b. system wear from day-to-day use

     c. changes in length of drive belts or chains due to use

     d. machine crashes

_____ 15. Which of the following is NOT true about global functions?

     a. they can be accessed by any program

     b. they save programmers time and effort

     c. they can be accessed by only one program

     d. some common global functions include open and close subroutines

_____ 16. The _____ determines the quality of a program.

     a. number of lines of code

     b. complexity of the code

     c. skill of the programmer

     d. all of these

_____ 17. In _____ programming, the programmer enters the positional data for each axis as well as all the motion, processing, and data gathering commands to create a program.

     a. level 1

     b. level 2

     c. level 3

     d. level 4

_____ 18. If there is no way to test a new program in manual before running it for the first time, you should _____.

     a. keep your hand near the E-stop

     b. make sure the work envelope is as clear as possible

     c. warn those nearby

     d. all of these

_____ 19. An example of _____ programming would be creating a new program, entering a string of points with proper motion labels, and testing the program for proper function.

    a. level 1

    b. level 2

    c. level 3

    d. level 4

_____ 20. We delete programs to _____.

    a. make room in the storage device and get rid of unused programs

    b. check for proper robot operation

    c. prevent them from being corrupted by continual use

    d. all of these

_____ 21. During programming, you need to look out for obstacles such as _____.

    a. fixture clamps

    b. tooling

    c. other equipment

    d. all of these

_____ 22. Level 4 programming systems are commonly used _____.

    a. in industry

    b. by hobby roboticists

    c. for BEAM research

    d. all of these

_____ 23. One area of modern robotics interested in level 1 programmed systems is the field of _____.

    a. AI robotics

    b. BEAM robotics

    c. industrial robotics

    d. none of these

_____ 24. The process of writing a program is _____ from robot to robot; it is the _____ from system to system.

    a. radically different, programming level that remains the same

    b. fundamentally the same, specifics or syntax that changes

    c. radically different, specifics or syntax that remains the same

    d. unique, code that remains the same

_____ 25. When you start writing a program, one of the first things you should do is _____.

    a. put the robot in the automatic or run mode

    b. check for or create the global subroutines as applicable

    c. change the tooling no matter how old it is

    d. delete the current running program to make room for the new program

Name: _____   Date: _____

Matching (1 point each)

Match the programming levels to the descriptions below.

a. Level 1
b. Level 2
c. Level 3

d. Level 4
e. Level 5

_____ **26.** No processor
_____ **27.** Simple point-to-point
_____ **28.** Point-to-point with AI
_____ **29.** Advanced point-to-point
_____ **30.** Direct position control

Matching (1 point each)

Match the terms to the definitions below.

a. And
b. Arcs
c. Call program/subroutine
d. Circular
e. Cycle time
f. End
g. Fixtures
h. Global function
i. If Then
j. Jump to
k. Local function
l. Macros
m. Math functions

n. Nor
o. Not
p. Or
q. Programming languages
r. Robot program
s. Singularity
t. Staging points
u. Subroutines
v. Swarm robotics
w. Third party
x. Wait
y. Weave
z. XOr

_____ **31.** The list of commands that run within the software of the robot controller and dictate the actions of the system based on the logic sorting routine created therein
_____ **32.** A portion of a circle
_____ **33.** A logic function that requires two or more separate events or data states to occur before the output of the function occurs
_____ **34.** Variables, subroutines, and other code or data accessible by any program you create on the robot
_____ **35.** Systems designed by someone other than the manufacturer or customer
_____ **36.** A sequence of instructions grouped together to perform an action that the main program accesses for repeated use
_____ **37.** This command stops the scanning of the program and triggers the system's normal end of program responses
_____ **38.** The opposite of the And command in that all the input conditions must be false before the output is triggered

_____ **39.** This command creates timed pauses in the program or has the robot wait for a specified set of conditions before continuing

_____ **40.** The rules governing how we enter the program so that the robot controller can understand the commands

_____ **41.** Commands that let you add, subtract, multiply, and divide in varying levels of complexity for data manipulation

_____ **42.** All input conditions must be false before the output occurs

_____ **43.** The field that focuses on ways to use a large number of simple robots to perform complex tasks

_____ **44.** A macro definition used to generate a sequence of instructions or other outputs

_____ **45.** A condition in robotics in which there is no clear-cut way for the robot to move between two points

_____ **46.** Works like the Or command except that only one of the conditions can be true for output to occur

_____ **47.** If at least one of two or more events happens, then the output of the function occurs

_____ **48.** The time it takes to complete the program

_____ **49.** Positions that get the robot close to the desired point but are a safe distance away, allowing for clearance and rapid movement

_____ **50.** This allows you to jump to a specified line or label in the program

_____ **51.** Data accessible by only one program

_____ **52.** Straight line or circular motion that moves from side to side in an angular fashion while the whole unit moves from one point to another

_____ **53.** This command calls up subroutines or other programs to help reduce the lines of code in a program

_____ **54.** Devices that hold parts in place for various industrial processes by clamping or holding the part in some manner

_____ **55.** Motion described by no less than three points to create arcs or circles

_____ **56.** An advanced logic filter that allows you to set a complex set of conditions to occur before a desired output function happens

Matching (1 point each)

Match the terms to the definitions below.

a. Continuous mode
b. Joint
c. Linear
d. NAnd
e. Step mode

_____ **57.** The manual mode in which the program runs until completed or you release a specific button

_____ **58.** Point-to-point motion in which all the axes involved move either as fast as they can or at the speed of the slowest axis, which may result in nonlinear robot motion

_____ **59.** Manual mode in which the robot executes one line of program code each time a specific button or button combination is pressed

_____ **60.** Motion in which the controller moves all the axes involved at set speeds to ensure straight-line motion

_____ **61.** The opposite of the And command in which all the inputs have to be false before the output is true

Short Answer (2 points each)

Write the answers to the following questions in the space provided.

62. What are some of the indicators that you are better off scrapping a program and starting over versus reworking the current program?

_____

_____

_____

_____

_____

_____

_____

_____

_____

63. What are some of the things that fall under the mapping question "Are there any conditions or other factors I need to consider in the process?"

_____

_____

_____

_____

_____

_____

_____

_____

_____

64. What are the drawbacks to writing a level 3 program on the teach pendant?

_____

_____

_____

_____

_____

_____

_____

_____

_____

65. What is the difference between creating a level 4 program with a teach pendant and creating a level 5 program on a robot with advanced teaching function?

_____

_____

_____

_____

_____

_____

_____

_____

_____

_____

_____

_____

66. What are some of the things we are looking for during the process of ensuring the program is logical, and what could happen if we skip this step?

_____

_____

_____

_____

_____

_____

_____

_____

_____

_____

_____

_____

67. What are a couple of the positive things about robot programs?

_____

_____

_____

_____

_____

_____

_____

_____

_____

68. What do you need to do to create a Weave command?

_____
_____
_____
_____
_____
_____
_____
_____
_____
_____
_____
_____

69. What is the potential problem with naming a new program the same as a program already in use?

_____
_____
_____
_____
_____
_____
_____
_____
_____
_____
_____

70. How do we avoid obstacles between two points when using linear motion?

_____
_____
_____
_____
_____
_____
_____
_____
_____
_____
_____
_____

71. What should you do during the first step of the program planning process?

_____

_____

_____

_____

_____

_____

_____

_____

_____

_____

_____

_____

Essay Assignment (12 points)

72. In Chapter 9, we discussed programming in general, so here is your chance to drill down into the specifics of programming the system(s) in your classroom. (Assuming you have robots to work with.) Take the time you need to answer the following questions about the lab robots you are working with. What is the programming language level and thus your responsibility in programming the robot? What is the normal method for creating and editing programs on the robot? What motion types do you have access to? What are the testing options once you write a new program, and how do you do so? Make sure you completely answer each question, since the more you know about the system, the easier it will be to create a program for it.

RUBRIC	1	2	3	4	Points Earned
CONTENT	The essay has only one of the following for the lab robot(s): programming language level with student responsibilities, normal procedure for creating and editing programs, motion types, and manual testing options with procedures.	The essay is missing two of the following for the lab robot(s): programming language level with student responsibilities, normal procedure for creating and editing programs, motion types, and manual testing options with procedures.	The essay is missing only one of the following for the lab robot(s): programming language level with student responsibilities, normal procedure for creating and editing programs, motion types, and manual testing options with procedures.	The essay details the following for the lab robot(s): programming language level with student responsibilities, normal procedure for creating and editing programs, motion types, and manual testing options with procedures.	
ORGANIZATION	There is no organization of the material.	There is some organization of the material, but it is still difficult to follow.	The work has a clear structure, but some of the organization interferes with clarity.	The work is clear and concise.	
GRAMMAR	There are four or more spelling, punctuation, or other grammar mistakes.	There are two or three spelling, punctuation, or other grammar mistakes.	There is one spelling, punctuation, or other grammar mistake.	Spelling, punctuation, and grammar are all correct.	

Research Assignment (12 points)

73. Using what you have learned in this book and other sources such as the library, Internet, or materials in your classroom, find an example of a robot that fits each of the programming language levels. For each robot you pick, detail what the robot is known as, the programming language level, when it was first created, who created it, and a brief description of its operation. As usual, list any resources you use per your instructor's preferred format.

RUBRIC	1	2	3	4	Points Earned
CONTENT	The report has only one of the following: a robot for each of the five programming language levels with name, language level, when it was created, who created it, a brief description of the operation, and a list of references.	The report is missing two of the following: a robot for each of the five programming language levels with name, language level, when it was created, who created it, a brief description of the operation, and a list of references.	The report is only missing one of the following: a robot for each of the five programming language levels with name, language level, when it was created, who created it, a brief description of the operation, and a list of references.	The report lists a robot for each of the five programming language levels with name, language level, when it was created, who created it, a brief description of the operation, and a list of references.	
ORGANIZATION	There is no organization of the material.	There is some organization of the material, but it is still difficult to follow.	The work has a clear structure, but some of the organization interferes with clarity.	The work is clear and concise.	
GRAMMAR	There are four or more spelling, punctuation, or other grammar mistakes.	There are two or three spelling, punctuation, or other grammar mistakes.	There is one spelling, punctuation, or other grammar mistake.	Spelling, punctuation, and grammar are all correct.	

Suggested Lab (value assigned by instructor)

74. Here is your chance to build a complex robot that you can use for multiple programming projects and then modify as your knowledge grows. Find the Spider Bot lab instruction in the activities manual and use this to create your own robot. This robot is a great platform for future experiments and programming exercises, and it is easy to customize with the equipment you have on hand. This lab will take a bit of time, especially if you solder together your own Arduino controller instead of using one ready to go, so make sure you plan accordingly. You may want to read the lab and glean the basics of operation so that you can make desired modifications as you build your robot. Perhaps you want to add the electromagnet to this build. Maybe you want to use an ultrasonic sensor instead of limit switches. Even if you do not add any of your own touches during the original build, the beauty of this system is that you can easily make changes later and see what happens. In fact, you can leave all the hardware as it is and make a few changes in the software to change the operation of the system and create a completely different behavior. I encourage you to play with the code, try some different things, and see what you can learn.

Use the provided lab form to assist with reporting your lab activities, unless directed otherwise by your instructor.

Spider Bot Lab Form

Lab Description

In the space below, describe the purpose of the lab and the equipment involved.

_____

_____

_____

_____

_____

_____

_____

_____

_____

_____

Lab Execution

In the space provided, detail the steps you took to perform the lab. Make sure to include any troubleshooting steps performed.

_____

_____

_____

_____

_____

_____

_____

_____

_____

_____

Observations

Record your observations about the system's performance here, including both the expected and unexpected.

_____

_____

_____

_____

_____

_____

_____

_____

_____

_____

Conclusions

What conclusions or statements can you make about the robot based on your observations and any data gathered during the course of the lab?

_____

_____

_____

_____

_____

_____

_____

_____

_____

_____

Name: _____ Date: _____

Score: _____ Text pages 238–277

ACTIVITIES

Multiple Choice (1 point each)

Identify the choice that best completes the statement or answers the question.

_____ 1. Which of the following is NOT true about equipment manuals?

    a. There is no set standard for information and format.

    b. You may find manuals that are similar to one other.

    c. Some manuals contain so little information that they are of little help.

    d. Manuals follow a standardized format.

_____ 2. What is the decimal equivalent of the binary number 111110?

    a. 63

    b. 31

    c. 62

    d. 52

_____ 3. We use _____ to set how many pulses an encoder has, limits for motor speed and axes movement, and other important variables.

    a. programs

    b. alarm codes

    c. parameters

    d. offsets

_____ 4. Which of the following is NOT something you check for when doing a continuity check?

    a. blown fuses

    b. bad connections

    c. voltage level

    d. damaged wires

_____ 5. Wiring diagrams show _____.

    a. components with no regard for placement

    b. components in their energized state

    c. the connections between components

    d. all of these

_____ **6.** Spending hours gathering data on the machine that does not tell you anything new about the system or point you in the direction of the problem falls into the realm of _____.

    a. good research

    b. busy work

    c. basic troubleshooting

    d. fixing the problem

_____ **7.** A good way to give the data you have gathered about a machine context without using manuals or machine literature is to _____.

    a. compare the data to an identical piece of equipment that is working

    b. compare the data to a different brand of machine that performs the same basic function

    c. compare the data to a machine from the same manufacturer that performs a different task

    d. all of these

_____ **8.** In the modern world, it is easy to gather large amounts of data from the equipment on top of the information available from resources such as manuals and the Internet. With this in mind, how do we filter all this information to make sure we focus on what is applicable to the problem at hand?

    a. by focusing on the information from the piece of equipment exclusively

    b. by focusing on the information found in the various resources such as manuals exclusively

    c. by using what we know about the machine and the problem we are working to correct

    d. there is no good way to filter this information

_____ **9.** Which of the following is NOT true when it comes to knowing the normal operation of a piece of equipment and the troubleshooting process?

    a. This allows the troubleshooter to determine how the faulted condition differs from normal operation.

    b. This allows the troubleshooter to find the conditions that are missing.

    c. It makes the job of troubleshooting easier.

    d. None of these.

_____ **10.** What is the decimal equivalent of the octal number 731?

    a. 734

    b. 374

    c. 537

    d. 473

_____ **11.** A high resistance reading on a fuse in a system _____.

    a. may be due to the circuit wiring

    b. may indicated a blown fuse

    c. is a good indicator you should pull the fuse and check it out of its holder

    d. all of these

_____ **12.** Most companies consider calling the manufacturer's technical support a last resort because _____.

    a. there is usually a cost involved

    b. the company may have to wait days for a technician to arrive if needed

    c. the hours of operation and hours that the technical support line is available do not match

    d. all of these

_____ 13. Diagrams of the equipment, technical instructions, and troubleshooting tests are all information we can get from _____.
    a. the robot's manuals
    b. the operator
    c. the alarm code
    d. personal observation

_____ 14. Which of the following would NOT be a good fit for the brainstorming technique?
    a. You run across a problem that no one has seen before.
    b. The system has one specific alarm.
    c. The system has an intermittent issue.
    d. The system has a large list of alarms that seem unrelated.

_____ 15. What is the decimal equivalent of the hexadecimal number 1A73F?
    a. 13,531
    b. 246,798
    c. 8,351
    d. 108,351

_____ 16. What is the octal equivalent of 1,833?
    a. 3341
    b. 3451
    c. 3541
    d. 3311

_____ 17. What is the hexadecimal number for 1,783,293?
    a. 7B356D
    b. 1B35FD
    c. 9C35FD
    d. 1A34FD

_____ 18. Which of the following is NOT a way we determine which test to perform first when there are several good choices for gathering information or determining the root cause of a problem?
    a. Pick ones that you have the equipment for.
    b. Start with those that require the least amount of time.
    c. The more complex the test, the higher the priority.
    d. If any cost is involved, go with the cheaper ones first.

_____ 19. What could happen if you replace a fuse or rest a circuit breaker without checking for the cause of the problem?
    a. Nothing; it is standard procedure to reset the breaker or replace the fuse before looking for any circuit issues.
    b. The circuit is likely to fault out again and possibly damage more components.
    c. The circuit may fault out again, but there is no danger of component damage.
    d. None of these.

_____ **20.** The _____ method asks specific questions or has actions to be taken that then direct the troubleshooter to another directive, depending on the outcome.

    a. flow chart

    b. dividing up the system

    c. tracking the power

    d. tracking the signal

_____ **21.** Schematics show _____.

    a. components in the energized state

    b. components to some degree as they appear

    c. how components are wired together

    d. the system in a logical fashion

_____ **22.** I/O use is limited by _____

    a. the number the system has available

    b. the voltage rating of the I/O

    c. the amperage rating of the I/O

    d. all of these

_____ **23.** What is the binary equivalent of 1974?

    a. 11110110110

    b. 11000110110

    c. 11110111110

    d. 11111110110

_____ **24.** When it comes to the troubleshooting process, looking for damage, determining what the machine was doing before it faulted out, and using your five senses falls under the _____ portion of the process

    a. information gathering

    b. action plan

    c. fixing the system

    d. testing

_____ **25.** Which of the following is NOT a benefit of asking others for help in the troubleshooting process?

    a. They may need some time to process the problem and need to do some checks of their own for understanding.

    b. They may offer a fresh point of view.

    c. Their experience may help in shedding light on the problem.

    d. Doing so can help you avoid tunnel vision.

Matching (1 point each)

Match each number on the picture to its description below.

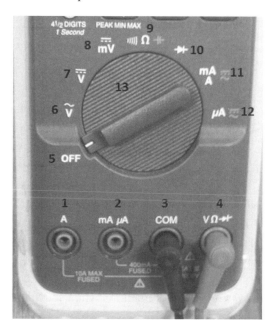

_____ **26.** This setting turns the meter off.

_____ **27.** This setting is used to check very small DC voltages.

_____ **28.** This setting is used to check micro amperage.

_____ **29.** This is where we plug the black lead for taking readings.

_____ **30.** This is how we designate the meter settings.

_____ **31.** This setting is for checking amperage in the milliamp or full amperage range.

_____ **32.** This is where we plug the red lead to take voltage, resistance, and electronic component testing.

_____ **33.** This is where we put the red lead to take small amperage readings under 1 A.

_____ **34.** This is the setting used to check AC voltage.

_____ **35.** This setting is for checking electronic components.

_____ **36.** This setting is used to check DC voltage.

_____ **37.** This setting is used to check capacitors, resistance, and continuity.

_____ **38.** This is where we plug the red lead to take amperage readings up to 10 A.

Matching (1 point each)

Match the term to the definition below.

a. Alphanumerical format
b. Analog
c. Arc flash
d. Binary
e. Bit
f. Brainstorming
g. Common connection
h. Continuity

i. Decimal numbers
j. Digital
k. Exploded view
l. Fuse bank
m. Flowchart
n. Hexadecimal
o. Intermittent
p. Legend

_____ **39.** Letters and numbers

_____ **40.** Data with a range of values based on the sensor or device

_____ **41.** Multiple fuses concentrated in one area for ease of wiring, checking, and replacement

_____ **42.** Unbroken electrical connection between two points

_____ **43.** Assemblies of complex systems drawn in such a fashion as to show all the parts that make up the assembly and how they fit together

_____ **44.** A condition in which a system will work fine for a while and then fault out without any clear-cut answer as to why

_____ **45.** Connections used to provide power or signal to NC and NO contacts

_____ **46.** A group of boxes with either a question or a task to perform with directions on where to go next

_____ **47.** A group of people getting together to come up with a variety of ideas about a topic or problem

_____ **48.** A base-2 number system with 0 or 1 for each position

_____ **49.** The base-10 numbering system that most of us are familiar with

_____ **50.** A base-16 numbering system that uses digits 0–9 and letters A–F for each position

_____ **51.** The smallest unit data can be broken into

_____ **52.** An electrical catastrophe in which two main legs of a system short out and draw enough amperage to create an electrical explosion complete with pressure waves, shrapnel, vaporized metal, and enough heat to instantly cause second- and third-degree burns

_____ **53.** Pictures of the symbols used in a drawing with details of what each symbol represents

_____ **54.** Data with a value of 0 or 1

Matching (1 point each)

Match the terms to the definitions below.

a. Live
b. Manuals
c. Mastering
d. Multiple failures
e. Octal
f. Parameters
g. PLC
h. Rungs
i. Schematics

j. Short circuit
k. Subchart
l. Tracking the power
m. Troubleshooter
n. Troubleshooting
o. Tunnel vision
p. VFD
q. Wiring diagrams

_____ 55. The specific pieces of data needed to run equipment
_____ 56. Electrically charged equipment
_____ 57. Programmable logic controller
_____ 58. A skilled person who determines the cause of a fault and makes the corrections necessary to get the system or process working properly once more
_____ 59. The books that come with a piece of equipment or system that give you deeper details on how it works
_____ 60. Variable frequency drive
_____ 61. When a person focuses on a specific way of thinking or seeing the problem and ignores everything else
_____ 62. When more than one component is the core cause of a problem
_____ 63. The process of setting or defining the home/zero position for each moveable axis
_____ 64. The lines going horizontally across the page of a schematic
_____ 65. A chart that takes a portion of a larger chart and breaks it into greater detail
_____ 66. Drawing of the electrical flow or logic of a system with no regard to component placement or design
_____ 67. The process of verifying that the components that should have power do indeed have power
_____ 68. A base-8 numbering system that uses 0–7 to represent each position
_____ 69. A direct path between a charged wire/component and a ground, neutral, or other powered wire/component
_____ 70. The logical process of determining and correcting faults in a system or process
_____ 71. Drawings that show how the power moves through the system using components that are to some degree the same as the actual components of the system they represent

Matching (1 point each)

Match each placement in the sequence to the troubleshooting steps below.

a.	First	**d.**	Fourth	
b.	Second	**e.**	Fifth	
c.	Third			

_____ **72.** The problem or condition

_____ **73.** Executing the action plan

_____ **74.** Testing the system for proper operation

_____ **75.** Forming an action plan

_____ **76.** Information gathering phase

Matching (1 point each)

Match the types to the guidelines below.

a. Signal guideline

b. Power tracking guideline

c. Troubleshooting failure guideline

_____ **77.** Reevaluate the information, adding in what you have learned

_____ **78.** Fuses blow and circuit breakers trip for a reason

_____ **79.** Check the input and output tables

_____ **80.** Do not get frustrated

_____ **81.** Check component by component once you have found the trouble area

_____ **82.** Match the I/O to the component/function it represents

_____ **83.** If everything else is fine, check the signal interpretation device

_____ **84.** Try, try again

_____ **85.** Determine why the component is not working

Short Answer (2 points each)

Write the answers to the following questions in the space provided.

86. When is it useful to use the ohms setting versus the tone setting on the meter for continuity checks?

_____

_____

_____

_____

_____

_____

_____

_____

_____

_____

87. What can happen if you misdiagnose a problem because you checked for continuity between two unconnected points?

_____

_____

_____

_____

_____

_____

88. What do schematics show, and how do they convey this information?

_____

_____

_____

_____

_____

_____

_____

_____

_____

89. What is the benefit of exploded view drawings?

_____

_____

_____

_____

_____

_____

_____

_____

_____

_____

90. What is the rule of thumb for selecting the voltage level on a meter when you are unsure of the voltage level you are testing?

_____

_____

_____

_____

_____

_____

91. What are some of the job titles that require troubleshooting skills?

_____

_____

_____

_____

_____

92. What is the downside to tracking power by tracing the wiring?

_____

_____

_____

_____

_____

_____

_____

_____

93. Where are some of the places we can find information about the I/O of a system?

_____

_____

_____

_____

_____

_____

_____

_____

_____

_____

_____

_____

_____

_____

_____

_____

94. What are some of the questions to ask about the system's operation before it faulted out?

_____

_____

_____

_____

_____

_____

95. What diagnostic tool do most technicians consider necessary for troubleshooting, and what are some of the things it can test?

_____

_____

_____

_____

_____

_____

_____

_____

Schematic to Wiring diagram (5 points each)

Use the schematic and components below to create a wiring diagram.

96. Using what you learned in Chapter 10, number the schematic below and turn it into a wiring diagram using the provided components. It is highly recommended that you color-code your wires for the sake of clarity.

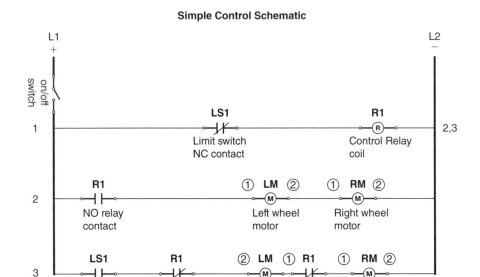

Simple Control Schematic

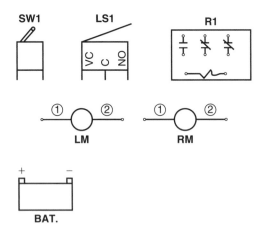

Essay Assignment (12 points)

97. In Chapter 10, you learned about the highly sought after skill of troubleshooting and in the "Finding a Solution" section, we talked about four different ways to sort information and fix a problem. Given the chapter's placement in the book, there is a high probability that you have already performed some troubleshooting of robotic systems and thus had a frame of reference, so here is your chance to share your troubleshooting story. Write an essay detailing a problem you faced, what you did to try to fix the problem, and if known, what the cause of the problem was. Once you have completed that portion of the essay, write a brief summary of how your method compared with the four methods mentioned in the chapter and how you would go about solving a similar problem in the future. Were there any similarities? Did you use a completely different method of solving the problem? Would you follow the same steps to solve the problem again? Would you try a completely different method to solve the problem now? These are the types of questions to answer or cover in your comparison summary. (If you do not have a robotic troubleshooting story to tell, talk with your instructor for further direction.)

RUBRIC	1	2	3	4	Points Earned
CONTENT	The essay has one of the following: a troubleshooting problem with corrective actions taken, root cause (if known), and comparison summary between student actions and four solution methods from the chapter.	The essay is missing two of the following: a troubleshooting problem with corrective actions taken, root cause (if known), and comparison summary between student actions and four solution methods from the chapter.	The essay is missing one of the following: a troubleshooting problem with corrective actions taken, root cause (if known), and comparison summary between student actions and four solution methods from the chapter.	The essay contains a troubleshooting problem with corrective actions taken, root cause (if known), and comparison summary between student actions and four solution methods from the chapter.	
ORGANIZATION	There is no organization of the material.	There is some organization of the material, but it is still difficult to follow.	The work has a clear structure, but some of the organization interferes with clarity.	The work is clear and concise.	
GRAMMAR	There are four or more spelling, punctuation, or other grammar mistakes.	There are two or three spelling, punctuation, or other grammar mistakes.	There is one spelling, punctuation, or other grammar mistake.	Spelling, punctuation, and grammar are all correct.	

Research Assignment (12 points)

98. In Chapter 10, it was mentioned that the Internet is a great resource for finding information to help with the troubleshooting process. Here is your chance to learn the process of finding that information without the stress of having a malfunctioning robot on your hands. Pick one of the three major robot brands—FANUC, ABB, or MOTOMAN—and perform some Internet research to find a make or model that is currently used by industry. Once you have picked your robot, perform several Internet searches to see what you can find of the following:

- Manuals
- Alarm codes
- Maintenance procedures
- Technical support
- Specifications

Once you have completed your searches, write a report detailing which robot you chose and why; whether you found each of the listed topics, with the website listed for those found; how helpful the information you found would be in the troubleshooting process; and how you found the various information. If you have trouble, refer back to the chapter for tips and tricks on searching the Internet for information. If you cannot find any information for the robot you chose, try picking another model and writing your report on that robot instead. The point of this exercise is to learn how to use the Internet to find helpful information, not to discover that you picked a robot without online resources.

RUBRIC	1	2	3	4	Points Earned
CONTENT	The report has only one of the following: which robot and why, the information found with website, how helpful the found information would be, and the process used to find the information.	The report is missing two of the following: which robot and why, the information found with website, how helpful the found information would be, and the process used to find the information.	The report is missing one of the following: which robot and why, the information found with website, how helpful the found information would be, and the process used to find the information.	The report details which robot and why, the information found with website, how helpful the found information would be, and the process used to find the information.	
ORGANIZATION	There is no organization of the material.	There is some organization of the material, but it is still difficult to follow.	The work has a clear structure, but some of the organization interferes with clarity.	The work is clear and concise.	
GRAMMAR	There are four or more spelling, punctuation, or other grammar mistakes.	There are two or three spelling, punctuation, or other grammar mistakes.	There is one spelling, punctuation, or other grammar mistake.	Spelling, punctuation, and grammar are all correct.	

Suggested Lab (value assigned by instructor)

99. Here is your chance to get some hands-on experience with troubleshooting. Talk with your instructor and have him or her designate a specific lab you completed previously. Do whatever is necessary to get the robot set up and ready to run, per the designated lab parameters. Once the robot is running correctly, have your instructor insert a fault condition while your lab group takes a break. Once your instructor gives you the go-ahead, use the information learned in Chapter 10 to determine the root cause of the problem and get the robot running normally once more. If your group gets lost along the way or cannot seem to find the root cause of the problem(s), ask your instructor for guidance. Also, do not forget the troubleshooting lessons you have likely already learned during this course. There is a good chance this is not the first time you have had to troubleshoot the robot, and your past experience is valuable information to keep in mind.

Use the provided lab form to assist with reporting your lab activities, unless directed otherwise by your instructor.

Troubleshooting Lab Form
Lab Description

In the space below, describe the purpose of the lab and the equipment involved.

_____
_____
_____
_____
_____
_____
_____
_____
_____
_____

Lab Execution

In the space provided, detail the steps you took to perform the lab. Make sure to include any troubleshooting steps performed.

_____
_____
_____
_____
_____
_____
_____
_____
_____
_____
_____
_____

Observations

Record your observations about the system's performance here, including both the expected and unexpected.

_____

_____

_____

_____

_____

_____

_____

_____

_____

_____

Conclusions

What conclusions or statements can you make about the robot based on your observations and any data gathered during the course of the lab?

_____

_____

_____

_____

_____

_____

_____

_____

_____

_____

| # Repairing the Robot

Name: _____ Date: _____

Score: _____ Text pages 278–299

ACTIVITIES

Multiple Choice (1 point each)

Identify the choice that best completes the statement or answers the question.

_____ 1. Using the tools you have handy instead of the right tools _____.

    a. can lead to damage to equipment as well as you

    b. is considered a normal practice

    c. saves time and money versus taking the time to find the right tools

    d. all of these

_____ 2. In cases in which you do not track down the root cause of the problem, _____.

    a. the fault was likely a fluke with no real cause

    b. it becomes the next technician's responsibility to find the issue

    c. make sure you perform all reasonable checks and pass along what you found to those you work with

    d. all of these

_____ 3. Which of the following is NOT true about stopping mid-repair to wait on parts?

    a. Parts may be used from the machine in other repairs.

    b. This is considered a preferred way to perform maintenance.

    c. Nuts, bolts, and other hardware may be lost.

    d. The person who finishes the work may be someone other than who started the task.

_____ 4. Which of the following is NOT on the checklist of tasks once you have finished a repair?

    a. Take a break and fill out all postrepair paperwork before you forget anything.

    b. Make sure everyone is clear and all the covers are in place.

    c. Check for any alarms or unusual action by the system.

    d. Start the robot in normal operation.

_____ 5. A machine has reached a zero energy state when _____.

    a. we lock out the electrical power

    b. we drain off all fluid power pressure

    c. we block the machine into a position in which it cannot move

    d. there is no kinetic or potential energy present

_____ 6. Finding a new or rebuilt part that is bad happens _____.

    a. rarely

    b. often

    c. never

    d. almost always

_____ 7. Which of the following is the least likely reason for new problems after you complete repairs?

    a. The system had multiple failures.

    b. You damaged something during the repair.

    c. The machine just happened to have a new failure right after the repair.

    d. You misdiagnosed the problem.

_____ 8. Which of the following is NOT one of the simple tests to check for proper system operation after repairs?

    a. moving each axis to check for proper operation

    b. creating a new program and checking for proper operation

    c. running a simple program such as the homing program

    d. stepping the robot through the program it faulted out in

_____ 9. If you make an offset to correct the issue and the system is better, but not fully fixed, what should you do?

    a. Remove the offset you made to get back to where you started.

    b. Make another offset in the same manner as the last.

    c. Make another offset opposite the last offset.

    d. Look for a new root cause of the problem.

_____ 10. Which of the following is NOT one of the arguments in favor of preventative maintenance?

    a. PMs can be scheduled.

    b. PMs save money.

    c. PMs may call for the replacement of parts that are currently working normally.

    d. Failure to perform PMs could void the machine warranty.

_____ 11. How do we determine when to change out components for preventative maintenance?

    a. by the cost of the part; the more expensive, the longer it lasts

    b. by the complexity of the part; the more complex, the sooner we should change it

    c. by analyzing run time data for trends or patterns of failure

    d. none of these

_____ 12. When looking for a place to set parts removed from a machine, you should avoid _____.

    a. areas where people are working

    b. areas near openings or grates

    c. areas where the parts could fall on you or others

    d. all of these

_____ 13. Before you begin repairs, it is a good idea to have _____ on hand.

    a. manuals, schematics, and diagrams

    b. tools to work on the equipment

    c. replacement parts

    d. all of these

_____ **14.** Who is legally authorized to remove a lock from the lockout device?

      a. supervisors only

      b. only the person who put the lock on

      c. anyone who has the key

      d. the person who put the lock on the lockout and his or her supervisor under specific conditions

_____ **15.** For machines that use large capacitors, the average safe discharge time is _____.

      a. 30 seconds to 1 minute

      b. 1–3 minutes

      c. 2–4 minutes

      d. 5–10 minutes

Matching (1 point each)

Match each frequency to the PM tasks below.

   a. Daily

   b. Monthly

   c. Quarterly

   d. Semiannual

   e. Annual

_____ **16.** Tightening electrical connections

_____ **17.** Greasing bearings

_____ **18.** Checking for leaks

_____ **19.** Replacing batteries

_____ **20.** Replacing filters

_____ **21.** Replacing the hydraulic oil

_____ **22.** Checking the level of the hydraulic oil tank

_____ **23.** Checking the robot for damage

_____ **24.** Checking for loose grounds on welding robots

Matching (1 point each)

Match the examples below to the category they would fall under.

   a. Parts swapping

   b. Repair

_____ **25.** Determining which parts of the system connect to the blown components on a drive card

_____ **26.** Tracking down a loose wire in the system

_____ **27.** Replacing the fuse and seeing what happens

_____ **28.** Analyzing the operation of the robot after correcting the issue that faulted the system out

_____ **29.** Replacing the overload on the motor starter and turning the system back on

Matching (1 point each)

Match the tasks below to the system you would perform them on.

  a. Hydraulic system
  b. Pneumatic system
  c. Electrical system

_____ **30.** Verifying muffler operation
_____ **31.** Filling the lubricator
_____ **32.** Verifying accumulator charge
_____ **33.** Tightening terminal connections
_____ **34.** Replacing the oil
_____ **35.** Changing batteries
_____ **36.** Checking the oil for water or dirt
_____ **37.** Replacing the wires inside the robot arm
_____ **38.** Cleaning and draining the reservoir

Matching (1 point each)

Match each task to its number in the LOTO sequence.

  a. Step one   **e.** Step five
  b. Step two   **f.** Step six
  c. Step three   **g.** Step seven
  d. Step four   **h.** Step eight

_____ **39.** Verify the machine is in a zero energy state.
_____ **40.** Perform repairs.
_____ **41.** Turn off or remove all external power supplies and lock them in the off position using lockout devices and a lock with your name on it.
_____ **42.** Remove all tools and any blocking devices or other items you added to the machine for safety reasons.
_____ **43.** Stop the machine cycle, if necessary.
_____ **44.** Return power to the equipment.
_____ **45.** Notify affected individuals you are about to shut the machine down.
_____ **46.** Place appropriate tags such as "Under Maintenance" on the machine.

Matching (1 point each)

Match the terms to the definitions below.

  a. Accumulators   **e.** Preventative maintenance
  b. Lockouts   **f.** Reseat
  c. LOTO   **g.** Zero energy state
  d. PMs

_____ **47.** Lockout/tagout
_____ **48.** A common name for preventative maintenance in industry

_____ **49.** Devices used to hold the power source in the blocked or de-energized state

_____ **50.** The process of removing electronic cards from the controller and putting them back in

_____ **51.** The practice of changing out parts and doing repairs in a scheduled fashion before the equipment breaks down or quits working

_____ **52.** When a machine has no active or latent power

_____ **53.** Devices used to store hydraulic pressure and then release it back into the system as needed

Matching (1 point each)

Match the tasks below the postrepair step they fall under.

a. Put everything back in its place

b. Finish the paperwork

c. Deal with the parts used

_____ **54.** Making notes in your technician's journal

_____ **55.** Completing your portion of the work order for repair

_____ **56.** Cleaning the tools you used and putting them back in the workbox

_____ **57.** Returning torque wrenches to their storage cabinet

_____ **58.** Ordering replacement parts

_____ **59.** Scraping out a damaged motor

Short Answer (2 points each)

Write the answers to the following questions in the space provided.

60. What are some situations to avoid during team lifts?

_____

_____

_____

_____

_____

_____

_____

_____

_____

61. What are the two possibilities when a system is worse after your repair efforts than when you started?

_____

_____

_____

_____

_____

_____

62. How do you ensure your safety in the lockout process?

_____

_____

_____

_____

_____

_____

63. What kinds of things do covers on the equipment protect us from?

_____

_____

_____

_____

_____

_____

64. What would be a couple of valid arguments against trying a new method of repair?

_____

_____

_____

_____

_____

_____

Essay Assignment (12 points)

65. One of the very important things we discussed in Chapter 11 is the process of lockout/tagout and putting equipment into a zero energy state. Using what you have learned in this chapter and what you know about the robots in your lab, write an essay on the specific process of rendering your robot safe to work on. Make sure to detail what is necessary for each of the eight steps of lockout for your lab robot(s). If you are working with systems like VEX or the LEGO NXT, there may be steps in the eight-part process of lockout that do not apply, and for those you need to detail WHY they are not applicable.

RUBRIC	1	2	3	4	Points Earned
CONTENT	The essay has at least two of the eight steps of the lockout process for a specific robot used in the classroom lab.	The essay is missing three or four of the eight steps of the lockout process for a specific robot used in the classroom lab.	The essay is missing no more than two of the eight steps of the lockout process for a specific robot used in the classroom lab.	The essay contains all eight steps of the lockout process for a specific robot used in the classroom lab.	
ORGANIZATION	There is no organization of the material.	There is some organization of the material, but it is still difficult to follow.	The work has a clear structure, but some of the organization interferes with clarity.	The work is clear and concise.	
GRAMMAR	There are four or more spelling, punctuation, or other grammar mistakes.	There are two or three spelling, punctuation, or other grammar mistakes.	There is one spelling, punctuation, or other grammar mistake.	Spelling, punctuation, and grammar are all correct.	

Research Assignment (12 points)

66. In the beginning of Chapter 11, we discussed preventative maintenance. Using resources such as classroom literature, the Internet, or the library, find a list of recommended preventative maintenance tasks for a specific industrial robot. Once you have this information, write a report detailing the make and model of the robot along with any daily, monthly, semiannual, annual, and longer preventative maintenance tasks for the system. Once you have done this, note any similarities between your research and what we discussed in the chapter, any differences, and anything you feel should be added to the robot's PM list and why. List the resources, per your instructor's preferred format, where you found the information.

RUBRIC	1	2	3	4	Points Earned
CONTENT	The report has only one of the following: make and model of the robot with a complete list of PM tasks sorted by time interval, a comparison of similarities and differences between the system and those listed in the chapter, and a reference section.	The report is missing two of the following: make and model of the robot with a complete list of PM tasks sorted by time interval, a comparison of similarities and differences between the system and those listed in the chapter, and a reference section.	The report is missing only one of the following: make and model of the robot with a complete list of PM tasks sorted by time interval, a comparison of similarities and differences between the system and those listed in the chapter, and a reference section.	The report lists the make and model of the robot with a complete list of PM tasks sorted by time interval, a comparison of similarities and differences between the system and those listed in the chapter, additions to the list and why (optional), and a reference section.	
ORGANIZATION	There is no organization of the material.	There is some organization of the material, but it is still difficult to follow.	The work has a clear structure, but some of the organization interferes with clarity.	The work is clear and concise.	
GRAMMAR	There are four or more spelling, punctuation, or other grammar mistakes.	There are two or three spelling, punctuation, or other grammar mistakes.	There is one spelling, punctuation, or other grammar mistake.	Spelling, punctuation, and grammar are all correct.	

Suggested Lab (value assigned by instructor)

67. Here is your chance to get your hands dirty and do some preventative maintenance. Ask your instructor to assign you a specific robotic system and then either use the provided resources or do the necessary research to determine the various preventative maintenance tasks necessary. Once you have a game plan of what to do, clear the list with your instructor and perform the tasks. You may need to grease bearings, change batteries, tighten connections, remove covers, check belt or chain tensions, or a host of other activities to complete this lab. Try to perform as many of the tasks as you can to increase your understanding of what it takes to keep a robot in running condition. Be careful not to damage the robot during this lab, as the point is to keep the robot running, not create more repair work. Once you are done, have your instructor verify your robot is ready to run and that you have performed the preventative maintenance satisfactorily.

Use the provided lab form to assist with reporting your lab activities, unless directed otherwise by your instructor.

Preventative Maintenance
Lab Form
Lab Description

In the space below, describe the purpose of the lab and the equipment involved.

_____
_____
_____
_____
_____
_____
_____
_____
_____
_____

Lab Execution

In the space provided, detail the steps you took to perform the lab. Make sure to include any troubleshooting steps performed.

_____
_____
_____
_____
_____
_____
_____
_____
_____
_____

Observations

Record your observations about the system's performance here, including both the expected and unexpected.

_____

_____

_____

_____

_____

_____

_____

_____

_____

_____

Conclusions

What conclusions or statements can you make about the robot based on your observations and any data gathered during the course of the lab?

_____

_____

_____

_____

_____

_____

_____

_____

_____

_____

CHAPTER 12

Justifying the Use of a Robot

Name: _____ Date: _____

Score: _____ Text pages 300–317

ACTIVITIES

Multiple Choice (1 point each)

Identify the choice that best completes the statement or answers the question.

_____ 1. The common range of robot precision is often within _____.
    a. 1/4–1/32 in.
    b. 0.1–0.01 in.
    c. 0.01–0.005 in.
    d. 0.005–0.0003 in.

_____ 2. You are talking with coworkers about the use of robots in the workplace, and the topic of robots replacing workers comes up. Which of the following is NOT one of the recommended points to remember?
    a. If the business is not profitable, it will not stay in business.
    b. Many tasks are better suited to humans than robots.
    c. Robots require a support staff.
    d. One robot may replace several workers.

_____ 3. When doing research, it is important to remember that the robot or robots needed depends on _____.
    a. the budget available
    b. which is the most advanced system
    c. the goal of the research
    d. current price trends

_____ 4. Which of the following is a potential cost impact to the company should a work-related fatality occur?
    a. wrongful death lawsuit
    b. low worker moral
    c. OSHA fines
    d. all of these

_____ 5. Which of the following is NOT one of the four Ds of robotics?
    a. dangerous
    b. dodge
    c. dull
    d. difficult

_____ **6.** Which of the following is true about the costs associated with workers injured on the job?

    a. The cost is low and rarely worth worry

    b. Cost is a concern only in cases where there are medical expenses

    c. The cost can be thousands if not millions of dollars

    d. The worker is responsible for all costs associated with on-the-job injuries

_____ **7.** The down side to PPE is _____.

    a. if it fails, the worker is exposed to a hazard

    b. it can be uncomfortable for the wearer

    c. the cost it adds to the overall process

    d. all of these

_____ **8.** Which of the following is NOT a cost associated with building a unique and new robotic system?

    a. the cost of repurposed parts from older robots

    b. the cost of the parts used

    c. the cost of the controller

    d. the cost of replacement parts

_____ **9.** The normal desired ROI for a robotic system is _____.

    a. 30–60 days

    b. 8–12 months

    c. 2 years or less

    d. 3–5 years

_____ **10.** Which of the following is NOT considered a primary option when the cost of the robot outweighs the return?

    a. Continue to perform the task without the robot.

    b. Cut positions not associated with the robot.

    c. Find a cheaper robot.

    d. Figure out a new way to use the robot.

Matching (1 point each)

Determine whether the statement below applies more to robots or human workers.

    a. Robots

    b. Human workers

_____ **11.** Costs the company at least $7.25 per hour

_____ **12.** Sensors register damage

_____ **13.** Costs the company about $0.72 per hour

_____ **14.** Has to use PPE to handle hazardous materials

_____ **15.** We can replace damaged parts easily

_____ **16.** Can often handle hazardous material without the need for extra equipment

_____ **17.** Deep space exploration

_____ **18.** Performs the tasks in the same manner continuously
_____ **19.** Damage results in the need for medical care and recovery time
_____ **20.** Working in areas with high levels of radiation for extended periods of time
_____ **21.** Experiences pain due to damage
_____ **22.** Can be distracted by circumstances both at work and outside of work

Matching (1 point each)

Match the terms to the definitions below.

a.	Exoskeletons	**e.**	ROI
b.	Justifications	**f.**	Service robots
c.	OSHA	**g.**	Turnover
d.	PPE		

_____ **23.** The government organization that ensures everyone has a safe and healthy working environment
_____ **24.** Employees leaving the company
_____ **25.** A measure of the amount of time it takes a piece of equipment to pay for itself
_____ **26.** Robots that perform tasks that people cannot do for themselves or would rather not perform
_____ **27.** Reasonable reasons for using robots
_____ **28.** Various items worn on the body to negate the hazards of a work task or area
_____ **29.** Robotic systems worn over the body that enhance the wearer's stamina or, in many cases, give mobility back to those who have lost the use of their legs

Matching (1 point each)

Match each robot below to the category it would fall under.

a. General task
b. Personal care
c. Risk mitigation

_____ **30.** Surveillance drone
_____ **31.** Telepresence robot for a child with severe allergies
_____ **32.** Roomba robot
_____ **33.** ReWalk exoskeleton system
_____ **34.** Bomb disposal robot
_____ **35.** Robotic wheelchair
_____ **36.** Lawnbot
_____ **37.** Sea mine sweeper

Short Answer (2 points each)

Write the answers to the following questions in the space provided.

38. When it comes to hazardous materials, what is the benefit of having the robot perform the task versus a human worker?

_____

_____

_____

_____

_____

_____

_____

39. What is the common justification for personal use entertainment robots, and where does the ROI figure in?

_____

_____

_____

_____

_____

_____

_____

40. What is the main benefit to students related to investment in research robot systems?

_____

_____

_____

_____

_____

_____

_____

41. What is the benefit of the time a robot runs after it has paid for itself?

_____

_____

_____

_____

_____

_____

_____

● Problems (2 points each)

Answer each of the following questions in the space provided

42. What is the ROI for a robot that costs $150,000 new, which includes the cost of shipping and installation? For a cost savings, we will use $0.30 per part after subtracting the operating cost of the robot, at a rate of 100 parts per hour. Finally, we will use 2,040 hours per year, as this is the standard 40-hour workweek times 51 weeks. Remember, the robot does not get sick days, paid vacations, or holidays, but we do factor in a week of downtime due to the facility being closed for holidays.

_____

_____

_____

_____

_____

43. What is the ROI for a robot that costs $127,283 new, which includes the cost of shipping and installation? For a cost savings, we will use $1.12 per hour operating cost versus the hourly wage of our fictional worker at $15.75 per hour when benefits and wages are figured in. Finally, we will use 2,040 hours per year, as this is the standard 40-hour workweek times 51 weeks, which is the amount of time the plant is open.

_____

_____

_____

_____

_____

44. What is the ROI for a robot that costs $247,833 new, which includes the cost of shipping and installation? For a cost savings, we will use $0.13 per part after subtracting the operating cost of the robot, at a rate of 45 parts per hour. Finally, we will use 8,568 hours per year, as this facility runs 24/7 for 51 weeks of the year. Remember, the robot does not get sick days, paid vacations, or holidays, but we do factor in a week of downtime due to the facility being closed for holidays.

_____

_____

_____

_____

_____

45. What is the ROI for a robot that costs $247,283 new, which includes the cost of shipping and installation? For a cost savings, we will use $1.32 per hour operating cost versus the hourly wage of the two fictional workers it replaces at $13.75 per hour when benefits and wages are figured in. Finally, we will use 4,080 hours per year, as this is the standard 40-hour workweek times 51 weeks, which is the amount of time the plant is open, multiplied by two because the robot replaces two workers and thus works two shifts each day instead of just one.

_____

_____

_____

_____

_____

_____

46. The best of both worlds. The company you work for has just purchased a robot that costs $500,000, which includes shipping and setup. This robot, due to its efficiency with consumables, saves the company $0.25 for each part produced and processes 65 parts per hour of operation. The plant runs 24/7 for 51 weeks out of the year, and the robot is expected to run 8,528 of those hours each year. (The deduction in hours is due to manufacturer-specified preventative maintenance.) This system also replaces three workers from the various shifts who had a total compensation of $18.50 per hour at 2,040 hours per year. The system has a running cost of $1.75 per hour and replaces 6,120 hours of operator wages each year.

First, what is the yearly cost savings in part cost using this robot?

_____

_____

Second, what is the yearly cost savings in wages using this robot?

_____

_____

Third, how long does it take the robot to pay for itself given the two yearly cost savings you just figured?

_____

_____

Essay Assignment (12 points)

47. In Chapter 12, you learned how we justify the use of robots in various applications. Taking what you have learned in this chapter, write an essay to justify the purchase of another robotic system like one you have been using in class or a whole new system you feel should be added to enhance student learning. Your essay should include the make and model of the robot, the approximate price, some form of ROI argument, and a list of reasonable justifications for purchasing the system. For the ROI, you will need to look at other factors besides the robot making money, as very few classroom robots produce any kind of sellable product. Good angles might be to talk about the improvement to the learning process, allowing for more students to work on labs, expanding the program, or preparing the student for work in industrial fields, to name a few examples. Write this paper as if you are giving it to someone who has never been in your classroom, with all that would entail.

RUBRIC	1	2	3	4	Points Earned
CONTENT	The essay has one of the following: make and model of the robot, approximate price, ROI, and reasonable justifications, all written for someone who has never seen the classroom.	The essay is missing two of the following: make and model of the robot, approximate price, ROI, and reasonable justifications, all written for someone who has never seen the classroom.	The essay is missing one of the following: make and model of the robot, approximate price, ROI, and reasonable justifications, all written for someone who has never seen the classroom.	The essay contains all of the following: make and model of the robot, approximate price, ROI, and reasonable justifications, all written for someone who has never seen the classroom.	
ORGANIZATION	There is no organization of the material.	There is some organization of the material, but it is still difficult to follow.	The work has a clear structure, but some of the organization interferes with clarity.	The work is clear and concise.	
GRAMMAR	There are four or more spelling, punctuation, or other grammar mistakes.	There are two or three spelling, punctuation, or other grammar mistakes.	There is one spelling, punctuation, or other grammar mistake.	Spelling, punctuation, and grammar are all correct.	

Research Assignment (12 points)

48. In Chapter 12, we talked a fair amount about ROI. Using resources such as information you can find in the classroom, the library, or the Internet, complete a ROI assessment on an industrial robotic system.

To determine the price of the robot, try to find three different price quotes for the system and average the three. For the hourly cost of operation, try to find information from the manufacturer. If you cannot find the information anywhere, use the $0.72 common rate we talked about in the chapter.

For the cost savings, find the average wage for three different factories that might use the robotic system and average the three for the base wage. Do not worry about trying to figure in fringe benefits or the like for the wage figure, as this information may be hard to find. When you get ready to crunch the numbers, use 2,040 hours per year for the worker time conversion and the initial ROI calculation. (See the example in the chapter if you do not remember how to do this.)

If the initial calculations show longer than two years for payback, determine how many workers total the robot would need to replace to pay for itself in two years or less. Finish the report with a brief summary of how and where you found your information (complete with references), any troubles you ran into while gathering data, and your observations about ROI.

RUBRIC	1	2	3	4	Points Earned
CONTENT	The report has one of the following: ROI calculations for two years or less, a summary of how the information was found with references, issues with gathering data, and the student's observations about ROI.	The report is missing two of the following: ROI calculations for two years or less, a summary of how the information was found with references, issues with gathering data, and the student's observations about ROI.	The report is missing one of the following: ROI calculations for two years or less, a summary of how the information was found with references, issues with gathering data, and the student's observations about ROI.	The report contains the ROI calculations for two years or less, a summary of how the information was found with references, issues with gathering data, and the student's observations about ROI.	
ORGANIZATION	There is no organization of the material.	There is some organization of the material, but it is still difficult to follow.	The work has a clear structure, but some of the organization interferes with clarity.	The work is clear and concise.	
GRAMMAR	There are four or more spelling, punctuation, or other grammar mistakes.	There are two or three spelling, punctuation, or other grammar mistakes.	There is one spelling, punctuation, or other grammar mistake.	Spelling, punctuation, and grammar are all correct.	

Suggested Lab (value assigned by instructor)

49. With most of my courses, when we reach the end of the book we are nearing the end of the time allotted for the course, and we are working to get everything finished up. With this in mind, ask your instructor if there are any lab activities you are missing or that you MUST finish. Take the time we would normally spend on a new lab to complete whatever lab(s) your instructor designates. To get proper credit, make sure you clearly indicate the lab you performed on the provided lab form. Do not forget there are blank lab forms at the back of the activities manual if you have more than one lab to complete.

If you do not have any labs to work on, this would be a great time to take all you have learned and see what you can come up with on your own. Maybe you want to make some modifications to the Box Robot. Perhaps you have some programming ideas for your version of the Spider Bot. Whatever the case may be, if you are caught up, this is the time to experiment on your own.

Use the provided lab form to assist with reporting your lab activities, unless directed otherwise by your instructor.

Classroom Lab Form

Lab Description

In the space below, describe the purpose of the lab and the equipment involved.

_____
_____
_____
_____
_____
_____
_____
_____
_____
_____

Lab Execution

In the space provided, detail the steps you took to perform the lab. Make sure to include any troubleshooting steps performed.

_____
_____
_____
_____
_____
_____
_____
_____
_____
_____

Observations

Record your observations about the system's performance here, including both the expected and unexpected.

_____
_____
_____
_____
_____
_____
_____
_____
_____
_____

Conclusions

What conclusions or statements can you make about the robot based on your observations and any data gathered during the course of the lab?

_____
_____
_____
_____
_____
_____
_____
_____
_____

Labs

About This Section

In this section, you will find the instructions needed to complete the suggested robotic build labs from the activities section. At the beginning of each lab, you will find a list of materials you need to complete the lab, and this is a good place to start. Once you have all the required materials, it is best to read before you actually start any of the steps. This will give you a clear idea of all the steps involved and give you a chance to ask your instructor any questions you might have. Once you begin the lab, use the provided pictures to help guide you through the steps.

Do not forget to complete either the lab form from the manual or the one provided by your instructor for each lab. Waiting until you have finished the lab to start filling out the lab form often leads to missing information. Regardless of when you fill the form out, make sure to complete each section fully and submit the form in a timely manner. If you are working as part of a group, ask your instructor if he or she wants individual lab forms or if a group lab form will work.

While the labs contained in this manual are low risk, that does not mean you can ignore the safety rules you have learned. Make sure to wear all necessary protective equipment and to exercise caution with the moving parts of the robot. Failure to follow the safety guidelines could lead to injury and loss of lab points. If you are completing these labs outside of the classroom, the safety rules still apply. Remember, the point of the labs is to learn about how the systems of the robot work, not how the robot can hurt you.

Vibro Bot Lab

Materials

- Toothbrush
- Pager motor
- Roll of electrical tape
- Small lengths of wire (22–26 AWG), preferably with at least one black and one red
- Button cell battery
- Glue gun or other adhesive system
- Needle-nose pliers
- Wire strippers (optional)

About the Lab

The Vibro Bot is a great place to start your robotic journey, as this simple system is controller free, highly flexible, and easy enough to construct that you can build a swarm of them for experimental purposes. These simple robots fall into the BEAM robotics kingdom and use the same principles that bring the Hexbug Nano to life.

Universities use these types of robots to study swarm robotic action, though many of those robots have specially printed bodies and are shrunk down to a much smaller scale than those we will be working with. After you build your first one, feel free to try changing things up and create your own experimental swarm of Vibro Bots.

When it comes to acquiring the parts for this build, feel free to dig around and scavenge parts where you can find them. I have found that dollar stores are a great place to pick up cheap toothbrushes. The wire I used for this project came from my workshop stash, but you could easily find what you need inside a battery-powered device that has quit working or you are ready to repurpose. The pager motor for this build I picked up at Electronic Goldmine, a great online resource to pick up various electronic components. Any small DC motor with an offset weight on the shaft would work, and you might find something in other things you have lying around or at your local electronics store, such as RadioShack. Once you have the basic components and something to attach them with, you are ready to begin.

Getting Started

To avoid issues along the way, gather everything you need for the project before you get started. I used a hot glue gun to stick my components together, but double-sided tape would likely work as well. I avoid using regular glue, as this creates a system that is difficult to change, and it takes a while to set up. If you go with the regular glue method, it may take longer to build as you wait for everything to dry, and you need to be extremely careful not to get any on the shaft of the motor.

Gathering everything together

Step One

To begin, we take the toothbrush and cut off the bristle end. If you want to keep the handle for another project, you are welcome to, but we need the bristle end for the body of our robot.

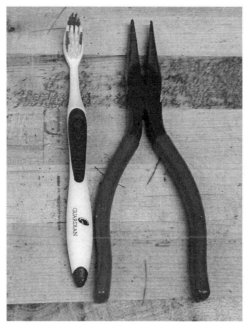

© Cengage Learning, 2016

Creating the robot body

Step Two

Next, we need to attach two wires to our pager motor to provide the power needed to run the motor. Depending on the motor you use, you may be able to skip this step if your motor is ready for power. The one I used had two terminals coming off the back, so I attached my wires there. I used a pair of wire strippers to remove the insulation necessary to make a good connection as well as remove insulation from the unused ends, as these will make our power connections later. I could have soldered the connections to the pager motor, but for this build, wrapping the wire around the post and crimping with the needle-nose pliers worked fine. I have seen builds that used hot glue to hold the wires in place, and that works as well, provided you keep a good wire to metal contact connection while the glue cools. If the glue gets between the two, it acts as an insulator and prevents the flow of power. Since I did not glue mine in place, I used two small pieces of electrical tape to maintain an insulating barrier between the two wires and posts. Failure to provide insulation between the two could short out the system and result in battery damage and improper or no operation.

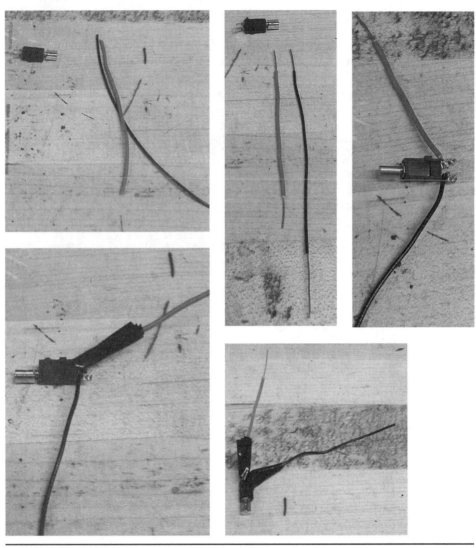

Attaching power wires to the pager motor

Step Three

Now that we have our motor prepped, it is time to attach it to the robot body, the toothbrush head we cut off in step one. I used the glue gun to complete this step. Because it dries quickly I can apply a large drop of glue to make the attachment and if something is not right, I can remove the glue and try again. I will admit that during one of the later steps, my motor did come loose because of this and I had to glue it down again, but I felt this was a small price to pay for the flexibility of using a glue gun.

I placed my pager motor near the front tip of the toothbrush head to make sure I had clearance for the battery and wires at the back. I also tilted the motor slightly to ensure the shaft and sling weight of the motor stayed clear of the glue and the problems that could cause. Feel free to experiment with the motor placement to see how it changes the motion of your Vibro Bot.

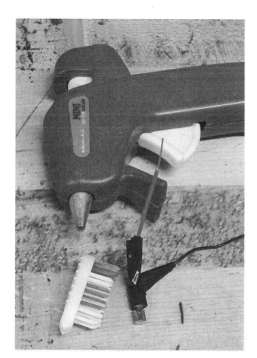
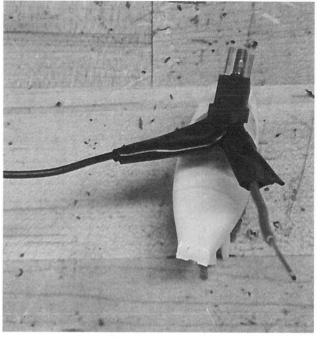

Attaching the motor to the robot body

Step Four

Next, we need to create our power supply. To power my robot, I wrapped the black wire around the remainder of the toothbrush handle, and this became the negative contact for the battery. To ensure good connection, I placed the button cell battery negative side down while pressing it firmly against the coiled wire and carefully applied a drop of glue to hold it in place. I waited long enough to make sure the glue was solid, 20 or 30 seconds I believe, and then carefully touched the red wire to the positive side of the button cell to make sure the motor turned. Once I confirmed motor operation, I added more glue to hold the battery in place and then cut a square of electrical tape to hold the red wire in place and act as the On switch for the robot.

Adding the power supply

Step Five

At this point, you are ready to test your creation and see how it moves. Mine kept falling over, and the first thing I noticed was the set of extra-long bristles toward the front of the toothbrush head. I took a pair of scissors and trimmed these down to help even out the balance of the system.

Troubleshooting

After trimming the bristles, I tried the bot again and found that it still wanted to fall on its side. I tried adjusting the wire positions to see if that would balance the weight but had no luck. So I took a third piece of wire and created two long legs for the robot that I routed between the motor and power supply for a quick test. This time, the robot took off bumping and exploring the world around it. I added a drop of hot glue to make the legs a more permanent attachment, and the Vibro Bot was ready to go. My cat especially liked this robot, as it is about the size of his mouse toys but able to skitter across the kitchen floor. This type of robot works best on smooth surfaces, and mine seemed to want to go under various appliances, so keep an eye on yours.

The finished Vibro Bot

Conclusion

Now that you have built the Vibro Bot, I encourage you to build others with differences in the design to see if it changes the performance. This is one of the core principles of BEAM robotics—create a simple robot, turn it loose, and see what happens. If you enjoyed this project, feel free to dig deeper into the realm of BEAM robotics, as you might find similar builds to pique your interest.

If this is an assigned lab, make sure you turn in the required paperwork. There are blank lab forms in the back of the activities manual to help with the reporting process if needed.

Electromagnet Lab

Materials

- Metal object such as a bolt, steel pin, or other small magnetic item
- Roll of electrical tape
- Roll of magnet wire (22–32 AWG)
- 9-V battery
- 9-V battery connector
- Switch
- Needle-nose pliers or wire strippers
- X-acto knife (optional)
- Multimeter (optional)

About the Lab

In Chapter 5, we talked about many different kinds of tooling for the robot, with the electromagnet being one of the types. In this lab, we are going to build a simple electromagnet that you could add to various robotic builds if you like. The one outlined here uses a simple manual switch to control current flow and thus the magnetic field, but with a bit of ingenuity you could use the processor of your robot to control the electromagnet. One caution I would like to add is avoid putting the electromagnet too close to the processor, as the magnetic field could cause problems with some processors and cause erratic operation of the robot.

When it comes to getting what you need for this lab, the metal core can be any handy metallic object that is magnetic and fits with the space you have to work with. I would recommend keeping the size somewhere between 1 and 3 inches long by no more than ½ to ¾ inch across. If you use too large of a core, the 9-V battery may not have enough electrical force and flow to generate a decent magnetic field. You can find the switch and magnet wire at various sites online or at electronics stores like RadioShack. This lab does not require a huge amount of magnet wire, so a small roll will be fine. Once you have all the materials listed gathered, you are ready to begin.

Getting Started

For this build, I used electrical tape to hold the coils of wire in place, but that is not the only option. I have seen builds that used spray on lacquer or clear fingernail polish to glue and hold the coils in place. This method is especially beneficial if you want to put more than one layer of coils on your magnet, which should increase the strength of the magnet. If you do not have electrical tape, you could use regular clear tape to hold the coils in place.

Gathering the materials

Step One

First thing you want to do is wrap your magnetic core in the magnet wire. Magnet wire is thin copper wire covered in a varnish-type insulation and is commonly used in speakers, motors, and other applications that generate magnetic fields to do work. If you have a multimeter handy and test two points along the magnet wire, you should not find any continuity.

When you start to wrap your core in the wire, make sure you leave a fairly long piece of wire sticking out to make connections later. I recommend leaving 6 to 8 inches, as you can always cut or fold up extra wire, but a lead that is too short could cause to all kinds of difficulties. When you make your wraps, you want to keep the wire tight around your core and slide the wraps close together. The more wraps and the tighter together they are, the better the magnetic field they will create. For the one pictured below, I stopped at one layer of wraps, but feel free to make two or three layers of wraps on yours if you like. For the multiple layer method, you need to use something to hold each layer in place so that you can start the next layer.

Winding the coil around the core

Step Two

Once you have the coil(s) of wire wrapped around the core, you need to use something to hold the coils in place. I used electrical tape for mine, but you could use regular tape, clear nail polish, spray lacquer, or anything of that nature. Since we have not found anything that insulates a magnetic field, you do not have to worry too much about what you use to hold the wire in place.

Once you have the coils secured in place, you can cut the wire leading from the core to the spool of magnet wire. Again, you want to leave around 6 to 8 inches of wire here so that you have plenty for connections later.

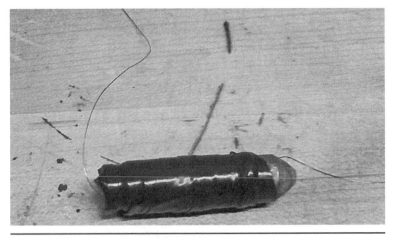
Holding the wire in place

Step Three

Next, we add in the battery connection and switch. The first thing we need to do is clear about an inch of each end of the magnet wire leads of the varnish insulation. This insulation is tougher than you might think, and the easiest way I have found to clean the wire is using my X-acto knife to scrape it off. You have to be careful with this method, as the wire is really thin, so it does not take a large amount of force to cut it. When you are done, you should notice a color difference between the insulated wire and the ends you just cleaned.

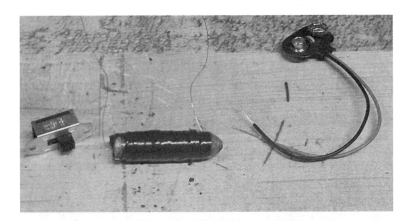

Cleaning the wire

Step Four

Now we are ready to attach the battery connection to the power supply. With battery-powered systems, we generally switch the red or positive lead, so connect the red battery lead to one of your switch contacts. The switch I used had a common and two control contacts, which is why I hooked my red wire to the middle contact. If your switch has only two contacts, then it really does not matter which one you hook the red lead to. If your switch has multiple contacts, use your meter on the continuity setting to find two contacts that have continuity with the switch on, verifying that these contacts lose continuity when the switch is off.

To make the wire connection, I wrapped the stripped end of the red lead around the contact and crimped it in place. If you have a soldering iron handy, you are more than welcome to make this a solder connection. I finished mine off with some electrical tape to prevent any shorting and help to keep the wire in place.

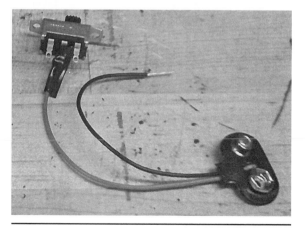

Connecting the positive battery lead to the switch

Step Five

Now we connect one of the magnet wire leads to the switch. It does not matter which one you pick. I used the same wrap, crimp, and electrical tape method from above for this connection as well.

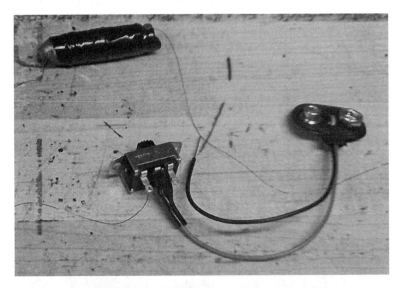

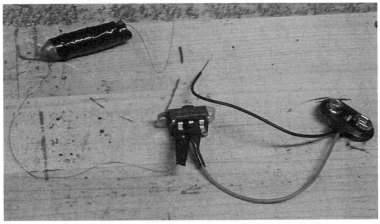

Connecting the electromagnet to the switch

Step Six

The next connection is the other magnet wire lead to the negative of the battery lead. For this connection, I simply wound the thinner magnet wire around the thicker battery lead and then used some electrical tape to hold it in place.

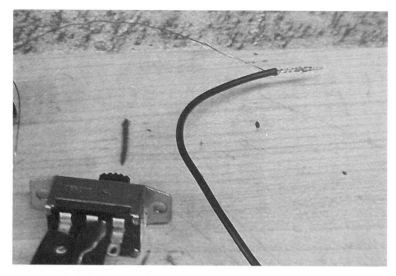

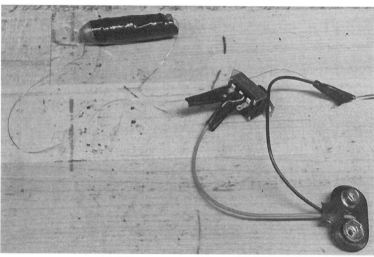

Connecting the negative terminal

Step Seven

You are ready to attach the battery, test your electromagnet, and see what it will do. Mine picked up a small screw decently, but it had to be pretty close. Using more wraps of wire should increase the magnetic field and increase the distance it will pull objects to it and how much it can hold. Remember to turn the switch on when you test your electromagnet. If it does not work, check to make sure you cleared all the insulation from the magnet wire ends and that you used the right set of contacts on your switch. This is where having a multimeter would come in very handy.

One caution—extended use of the electromagnet might heat up the wire due to the current flow. I recommend you carefully check the temperature of the electromagnet from time to time if you use it for extended periods to ensure it does not get too hot and cause problems.

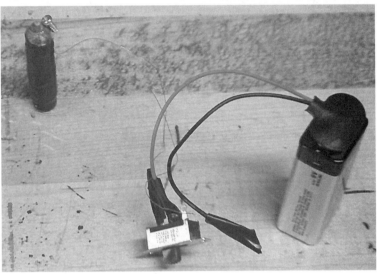

Ready to test

Conclusion

Now that you have a working electromagnet, you have another way for your robot to interact with the world. I encourage you to look at ways you might add this to an existing robot you are using in the classroom or perhaps one you built at home for fun. If you have a relay contact or similar on your robot, that can take the place of the switch and allow the robot to control the electromagnet in the same manner it might a gripper. The options for using this are limited only by your imagination and cunning.

If this is an assigned lab, make sure you turn in the required paperwork. There are blank lab forms in the back of the activities manual to help with the reporting process if needed.

Materials

- Toysmith TS4687 Box Robot kit
- AA battery
- Pair of scissors
- Small Philips screwdriver
- Ruler (optional)
- X-acto knife (optional)

About the Lab

The Box Robot is a fun robotic kit for the beginner that builds a simple little rover-type robot. This system has no sensors and thus does not interact with its environment, but it is a quick and fun way to learn the basics of loco-motion and has potential for modification as your robotic knowledge grows. I picked up my kit at a local Hobby Lobby, but I have found them online at sites such as Amazon and Maker Shed. When you get your kit, be careful with the box, as this is the body for your robot. The directions that come with the kit are straightforward, but I did find a few minor details that might be tricky if you have no experience putting together models. Without further ado, let us begin.

The Kit

The white wrapper around the box contains the instructions for building the kit as well as several pictures of ideas for building your robot. You will want to keep this handy as you put your robot together. Again, make sure you take care not to damage the box, as this is the body of the robot, and it is made of cardboard and thus vulnerable. When you open the kit, here is what you will see:

Inside the box

Step One

The first thing I do when working with kits of this nature is to set everything out and see what I have. If you look at the inventory picture on the next page, you will see two cardboard cutouts on the right that later become the hands of the robot and two on the left that are there for whatever you would like to make them into. The instructions in step one say to cut off the smaller tabs of the box, where the cardboard cutout pieces come from, but I left mine on until I determined if they needed to be removed. It turns out that for the upright configuration, the two tabs that hold the robot hand cutouts have to go, but the other two did not get in the way of the build. I left the attached portions of these tabs in place to add some stability to the box.

Inventory photo

Step Two

Next, we take the motor base, which is the large green piece with the two wires coming off it, and snap the shaft with the gear on it in the front portion. Make sure you center the shaft so that the gear has clearance to turn between the supports; otherwise, this could cause interference with the operation of the system.

Axel insertion

Step Three

The next step is to set the motor in its designated cradle, making sure we have a good positive connection between the worm gear on the motor and the gear on the axel shaft (see the pictures below). Make sure no metal screws or other small metallic items are stuck to the motor, as this is a permanent magnet DC motor and thus has a magnetized motor case. When I took mine out of the box, all the screws for the kit were stuck to it, so watch out for this. Also, make sure to place the motor in the cradle with the wires sticking up, not down.

What you should see when you place the motor correctly; notice how the teeth of the motor's worm gear meshes with the axel gear

Step Four

The next step is to make the motor connections. Another interesting thing about this kit is how they chose to make the power connections. There are two circular portions on the motor base ringed in metal, and this is where we make the power connections. Place the black connection from the motor and the battery area in the one on the left and the red wire from the motor and battery on the right. If you look at the next page, you can see the before and after pictures of making the connections as well as the two plastic pieces that hold the wires in place. The benefit of this system is twofold; one, there is no need to solder the connections together and two, this makes it easy to power any additions you might want to add down the road.

Motor connections, setup and completion

Step Five

With the motor connections made, we are now ready to put on the motor cover, which is the green piece with the curved trough on the front. This is the first time we use any of the screws in this kit, and hopefully you noticed that we have several screws that are all the same and one with a large built-in washer. Make sure you save the one unique screw for later and use a couple of the general screws to secure the motor cover. This keeps the motor in place and makes sure the system maintains good gear contact. When finished, your system should look like the picture below.

Motor cover installation

Step Six

The next step is to place the motor base assembly into your Box Robot, and this step is a bit unclear in the provided instructions. To do this, you need to find the flap that has the rectangular cutout and punch this out. Once you have the done this, align the On/Off switch with this cutout to help with orientation. This is a good time to pop out the two circular holes that align with the axel. Be careful not to damage the box as you pop out the cutouts. During the building of this kit, I had only one punchout that hung up, but it did create a blemish on the box. If you are confused, use the images below to guide you.

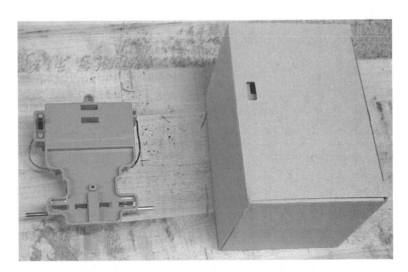

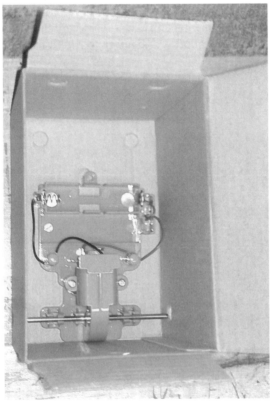

Inserting the motor assembly into the box

Step Seven

Next you are going to secure the base using one of the star-shaped green pieces and a screw that goes through the inside of the motor base, above where the battery goes. This aligns with the hole that becomes the nose on the upright robot, near the rectangle for the On/Off switch. Pay special attention to the fact that the green star piece has a flat on the backside that lines up with a matching flat in the motor base. Failure to align the two could cause misfit and the temptation to use a level of force that could damage the box. Take a moment to look at the photos below before you complete this step and keep in mind it should not take a large amount of force to make this connection. Once you have the green star piece seated, use one of the generic screws to secure it from the inside of the box.

 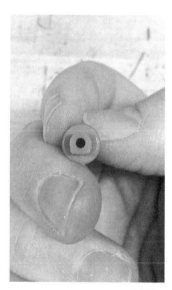

First connection to secure the motor assembly

Step Eight

Next, we put the wobbly wheels on the robot. They made these wheels in such a way that the flats do not quite line up, thus the wobbly description. You want to be careful with this step, as it takes a fair amount of force to push the wheels on the axel. I started each wheel and then pushed on both wheels at once using a slight back and forth rotation to work them onto the shaft. I opened the back of the robot box and made sure they were even while doing this. Again, the body of the robot is made of cardboard, so be careful not to damage the robot.

Wheel attachment

Step Nine

Now we are ready to use that unique screw to finish securing the motor assembly. We waited until after we put the wheels on so that the whole assembly could self-adjust as needed, but now it's time to lock it down. Punch out the small circular hole and thread in the unique screw with the built-in washer. Make sure you secure the screw but do not overtighten, as this can damage the cardboard body.

The unique screw as well as the area we thread it into

Step Ten

The next step in the instructions is to use the other green star piece and a green base piece to create a drag point for the upright robot. If you have decided to create the long robot rather than the tall robot, you can skip this step and save the pieces for later. The whole point is to create a low friction point for the robot to rest on instead of trying to drag the whole edge of the robot. The first motor assemble attachment provides this for the long robot. I created the tall robot, so I popped out the circle on the base and inserted the star piece into the screw base, taking note of the orientation of the flat. See the pictures below if you are having trouble with this step.

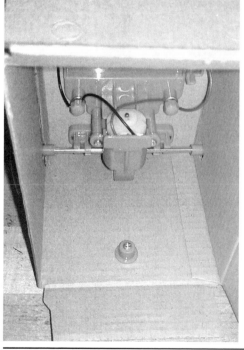

Low-friction point assembly

Step Eleven

Now it is time for the motor test. You need to insert the AA battery in the battery holder with the positive tip facing the red wire and the negative side against the spring. This kit has a healthy battery spring, and I found it took a fair bit of force to insert the battery, so again be careful with the box. Once you have the battery in, close up the box, flip the robot over, and turn it on. I held mine in my hand, as I just wanted to make sure the wheels turned, and this prevents any potential for accidental damage to the box.

If your wheels do not turn, make sure the battery you put in is good, make sure you did indeed turn the switch on, and check to make sure the wires are making good contact with the metal ring provided on the motor assembly base. If none of these fixes the system, start making sure the wires are firmly connected to the motor and the battery terminals. Though slim, there is always the chance something is wrong with your switch or motor as well, and this would require getting a replacement kit.

Test time

Step Twelve

Now it is time to give you robot arms, if you so desire. To do this you need to make slits in the box wherever you want the arms and then insert the tabs into this slit. I used an X-acto knife to make my slits and measured from the top of the box to make sure they were even. I found that after making the cut, wiggling the X-acto knife in the slit opened it up a bit and made it easier to get the tabs in. Once inserted, I bent the tabs over to hold the arms in place, as shown below.

Adding arms

Personal Touches

For my robot, I decided to add a spikey top, which I cut out of one of the spare side flap pieces removed in step one. I used the X-acto knife to cut out the design and create a tab to attach my addition. I also used the provided marker to color the spikes and make some designs on the arms and robot body. I used the same procedure for attaching the arms to attach my spikes, as seen below.

 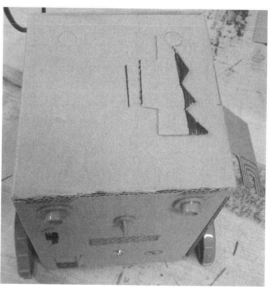

Adding some personal touches

Finishing Up

Once you have customized the box however you like, it is time to kick it on and let your robot run. The little creation runs quickly and has no real ability to turn, so keep this in mind as you turn it loose. Once you have completed the robot, it would make a fun base to work with, adding new things as your knowledge grows. You could add some LEDs to the eyes. Maybe you want to add a micro limit switch and rewire the system to make it reverse when it hits something. (Connect the motor red to the battery black and the motor black to the battery red to reverse the motor direction.) You might add a second drive system to turn the robot. The possibilities and choices are up to you.

How mine looked all finished

Conclusion

Now that you have some experience with drive systems, I encourage you to experiment with the Box Robot and whatever materials you have on hand to expand your knowledge. Many times in the classroom, we work with robots without seeing how the gears mesh up and what is going on under the covers. Simple kits like this allow you to get to the heart of a system and see how everything works together.

If this is an assigned lab, make sure you turn in the required paperwork. There are blank lab forms at the back of the activities manual to help with the reporting process if needed.

Obstacle Avoiding Robot Lab

Materials

- Obstacle Avoiding Robot kit made by ArTeC #095061
- Roll of electrical tape
- Roll of clear tape
- Needle-nose pliers
- Scissors or X-acto knife
- AA batteries
- Metric ruler (optional)

About the Lab

The Obstacle Avoiding Robot is another cardboard body robot, but unlike the other one outlined in this manual, this robot can react to its environment. With the addition of two micro limit switches and a way to trigger them, this robot has the ability to reverse direction when it runs into something in its path, thus creating a completely new operational capacity.

I picked this kit up from Maker Shed online, but you can find it at various online shopping sites such as Amazon or possibly places like Hobby Lobby or other hobby stores. You can replicate this kit on your own if you like, but by the time you track down and pay for all the parts, I think the cost works out about the same or possibly a bit higher than the kit.

The Kit

Unlike the Box Robot, the Obstacle Avoiding Robot does not use the shipping box for the robot's body, so damage to the outer box is not a big issue. Inside you will find the three cardboard body pieces, two arms, two limit switches, two motors, some wire, a switch, two battery holders, six red wire connectors, and a package with thin clear double-sided tape and thick white double-sided tape. I went ahead and laid everything out to make sure I had all the pieces before beginning, since it is easier to return an unused kit rather than a half-completed one.

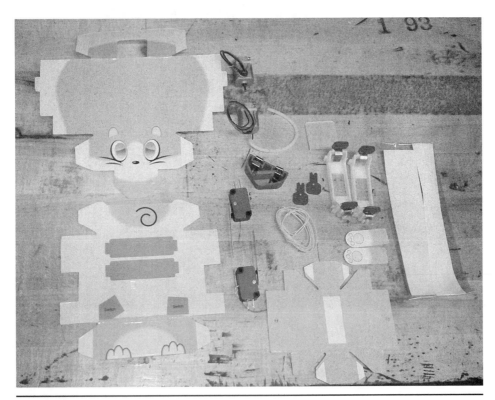

Inside the box

Step One

The first part of this build is to create the motor housing. For this step, you take body piece number 2 and fold it so that the white underbelly makes an area for the motors to go. I used about 1.5 cm of the clear double-sided tape on each tab to hold the assembly together, but you may need to add a bit more if yours tries to come apart. I had two full pieces of the clear double-sided tape left, so using a little extra here and there should not be a problem.

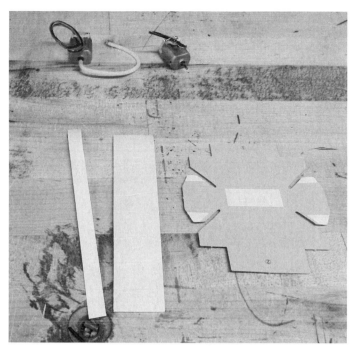

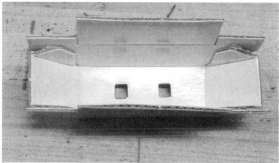

Preparing the motor assembly

Step Two

Next, you need to take something like a paper towel or napkin and wipe the flat side of the motors in preparation for the tape. Both of the motors in my kit looked clean, but after wiping them down, there was some oil on the paper towel. Once you have the motors prepped, apply a piece of the thicker white double-sided tape 1 cm wide by 2 cm long. I lined my pieces up with the shaft edge of the motors. For all the double-sided tape applications, I would expose one side of the tape, secure this, and then expose the second side of the double-sided tape. This prevents it from sticking to your fingers, the table, or anything else you do not want it to hold in place. The directions for attaching the motors instruct you to leave the motor shafts protruding slightly from the assembly, as this is how your robot will move later on. The closer you can get the two motors to protruding the same, the straighter your robot will run. Mine was off a bit, as I discovered when I saw the robot's tendency to veer right.

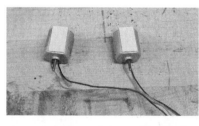

Attaching the motors

Step Three

Now you are ready to thread the motor wires through the provided holes in the back of the motor assembly and close everything up. I used one strip of the clear double-sided tape about 5 cm long at first, but this came open later on. I added another strip of the same tape, about 4 cm long, to help hold everything together. This worked for most of the build, but ultimately I wrapped the outside of the assembly in clear tape to add additional support (pictures of that can be found later in the lab). If I were doing this build over, I would go ahead and use the two pieces of the double-sided tape between the large flaps and then a couple of wraps of clear tape all the way around in the middle to help hold it all together.

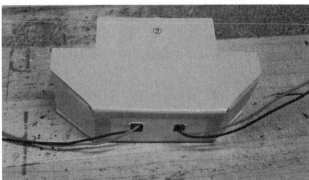

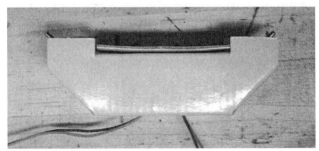

Closing up the motor assembly

Step Four

The next step is to put small pieces of the provided rubber hose onto the motor shafts. This is what acts as the wheels for our robot and how it gets traction. They sent a sizable piece of hose with my kit, so I have plenty to replace the "wheels" as they wear out. The instructions say to leave about a millimeter sticking over the end of the shaft and to leave another millimeter at the base. I measured my motors, and they had 8-mm shafts, so I just cut my pieces of the rubber hose 8 mm long and left the recommended millimeter of hose protruding from the shaft.

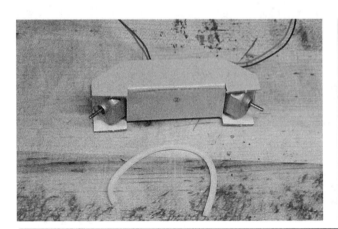
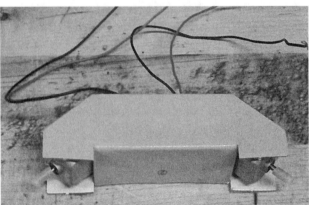

Putting the wheels on

Step Five

Next, we begin on the lower section of the robot body, and this is where everything else connects, so you will be working with this part of the robot for a while. First things first, you need to take body part number 3 and fold it into shape so that the gray shaded areas face up and the colored panels face out. There are several precut pieces on this part of the robot, and you can go ahead and knock these out, setting them aside. I used 1.5-cm pieces of the clear double-sided tape for the flaps on this piece as well and did not have any major problems with them coming loose.

 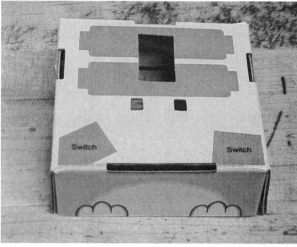

Prepping the robot base

Step Six

Now you are ready to attach the motor assembly to the robot base. I used three pieces of the clear double-sided tape, as pictured in the next page, to secure the motor assembly to the base. The clearance provided for the motor assembly is tight, so be careful that you do not damage the base or motor assembly during installation. Also, make sure the two square holes you ran the wires through in the motor base line up with the two square holes in the robot base. You need to run the motor wires through the two identical square holes in the robot base that line up with the motor assembly so that you can power the motors later. If you have questions, refer to the pictures in the next page.

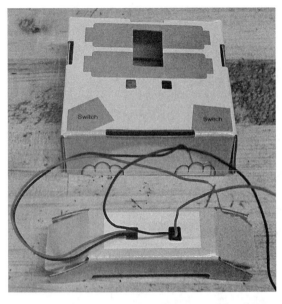
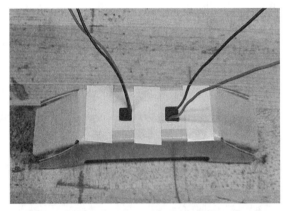
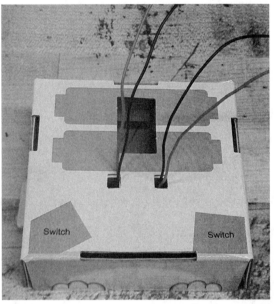

Joining the motor assembly to the robot base

Step Seven

The next step in the process is preparing the power supply. You need to get the two yellow battery holders and snap them together, as shown in the next page. Once that is done, you need to cut two pieces of the thick white double-sided tape 1 cm wide by 3 cm long. In case you have not measured yet, the strip of thick double-sided tape is 3 cm wide, so you need to measure only 1 cm down from the end and cut.

The instructions have you go ahead and prepare the six connection wires at this point before sticking your battery holder to the base, but I see no reason why you could not go ahead and put on the battery base before preparing the wires. The choice is yours. When you put the battery pack on the base, make sure you have a positive terminal connection in the upper left corner, with the switch portion of the base facing you. This makes sure you wire the system correctly and prevents your robot from running backward when you do not want it to.

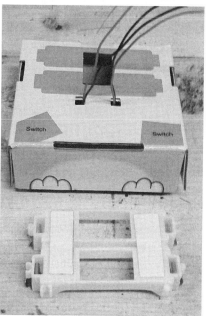 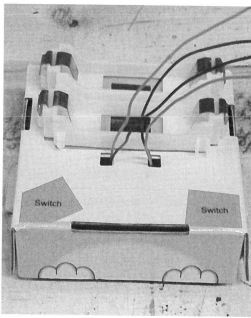

Preparing the battery pack

Step Eight

Now it is time to take the spare wire and cut it into six equal lengths approximately 20 cm long. After that, the instructions have you strip 3 cm of each end of the wires using your fingers. I admit that I was really skeptical about this, but I tried it as instructed and found it easy to strip the wires in this way.

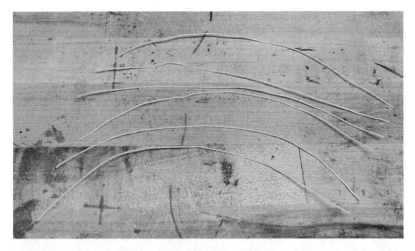

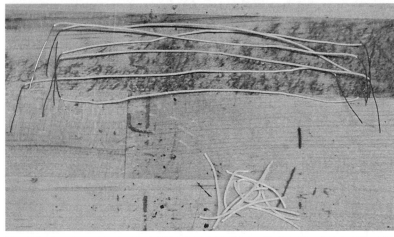

Cutting and striping the connection wires

Step Nine

The next step is the first of the motor connections to the battery supply and one of the hardest tasks in the whole build. First, you need to take one red motor lead and one spare yellow wire and wrap them carefully around the provided red plugs, securing the end of the wires around the two wires coming in.

Preparing the wire connection

Once you have made a red tab for each motor, it is time to place these in the battery holder on the left side as you face the base with the switch positions in the front. The red lead from the motor on the right goes in the left battery slot at the back of the battery pack. It takes a lot of force to get these plugs in once you add the wire, and the only thing I found that helped was to continuously rock the plug from side to side in the designated slot until it finally started to slide into place. The battery connections are a tight fit, and you will just have to work with this until you get them properly seated.

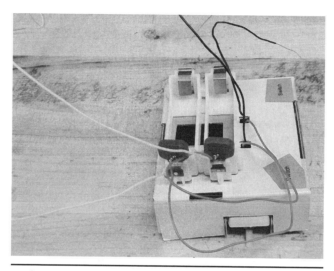

Making the first set of connections

Step Ten

Up next is prepping the two limit switches. You are going to take the remaining four yellow connection wires and put one wire on each of the contact terminals of the limit switches. Make sure to put these on the two connections coming out the back of the limit switch and NOT the common connection coming out the base or side opposite the lever arm of the switch. I wrapped the stripped ends of the wires around the contact post, crimped them in place, and then used a piece of electrical tape to ensure isolation. You should end up with something like the pictures below.

Getting the limit switches ready

Step Eleven

The last step in preparing the switches before attaching them to the base is to connect the blue motor lead to the common of the appropriate switch. The blue lead from the motor on the left of the base goes to the limit switch we will attach on the right and vice versa for the other motor. When properly connected, the two blue wires cross over each other. We then use two 1-cm by 1.5-cm pieces of the thick white double-sided tape to hold the switches in place. I put my tape on the base first and then stuck the limit switches in place, as this seemed the easier method to me. You want to try to orientate the switches in a similar manner, as these are the sensor arms of your robot. I then coiled the extra blue wire up to provide greater clearance for everything.

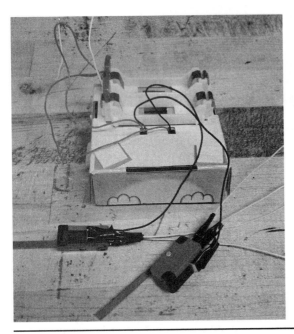

Wiring and attaching the limit switches

Step Twelve

The last step before the initial operation test is to make the connections from the limit switch contacts to the power supply. You will take the NC contact of the left limit switch and combine it with the NO contact of the right limit switch; once you have properly inserted them into a red tab, you will place this in the battery terminal closest to you on the right or the + terminal. Next you will take the NO contact from the left limit switch and the NC contact from the right limit switch and connect this to the back battery terminal on the right or the − terminal. When done, your circuit should look like this:

Connecting the limit switches to power

This is what creates the circuit that reverses the motor when the robot runs into something. With the switches in their normal state, the NO contacts do not make connection, and the NC contacts pass power. The red lead of the right motor is tied to the + terminal of one battery, and the blue lead of the right motor is tied to the left limit switch, which in turn has its NC contact tied to the positive terminal of the battery closest to the switches. The red lead of the left motor is tied to the – terminal of one battery, and the blue lead of the left motor is tied to the right limit switch, which in turn has its NC contact tied to the negative connection of the other battery. At first glance, this seems like faulty wiring and the motors should not run. In truth, they do not, until we connect the two yellow wires that go to our switch in a later step. This connects the positive of one battery to the negative of the other and allows the two motors to finish the circuit.

When we engage a limit switch, it changes the contacts, and this in turn reverses the polarity of the power supplied to the motor. With a reverse in power polarity, the motor runs in reverse, and thus the robot backs up or turns away from the obstacle. Another unique situation this wiring creates is the fact that the motors will turn anytime you have batteries in the system and press one of the limit switches. This is something I learned as I finished the build and more than once the motors started turning.

Speaking of testing, you are now ready to see if your motors turn. Place two batteries in the battery pack and then carefully cross the two yellow wires to see if your motors turn. I recommend holding the base in your hand to prevent the robot base from wandering off.

Ready to test

Step Thirteen

If the motors turned when tested and reversed when you pressed a limit switch, you are now ready to finish the build. The first thing we want to do at this point is attach the remaining two wires to the remaining two red wire tabs.

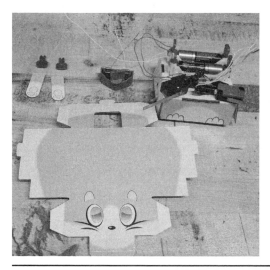

Preparing the switch connections

Step Fourteen

Next, we want to attach the green switch to the upper body half so that the triangular projection where the switch turns extends out the square hole in the back of the assembly. I folded up the back flap until it was square with the table in the position I believed the flap would end up in when finally assembled. I then placed the switch in its square cutout and positioned it so that I could engage the switch from the outside. I did this WITHOUT tape and made some alignment marks so that I would know where to put the switch once I had tape on it. Then I applied the thick double-sided tape to the switch, a strip 1 cm by 3 cm, and attached it to the upper half using my alignment marks.

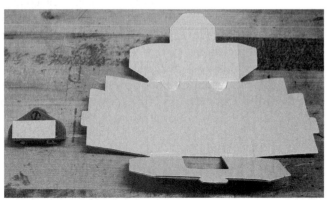

Aligning and attaching the switch

Step Fifteen

You are now ready to finish the folds and tape the tabs to turn the top half into a box shape. I used two pieces of the thin double tape 4.5 cm long and two 3-cm long pieces on the tabs.

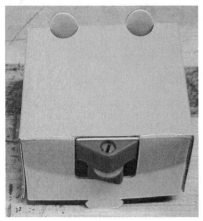

Getting the top ready

Step Sixteen

Now you are ready to make the two connections to the switch. These connections went in easily, as each red tab only had one wire and due to the way the circuit is set up, it really does not matter how the tabs are placed as long as you have one tab on each side of the switch.

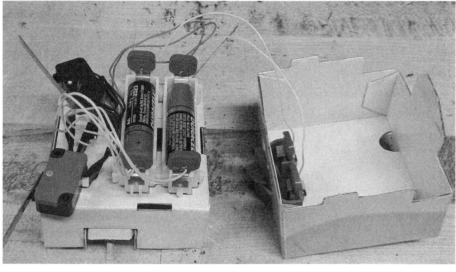

Making the switch connections

Step Seventeen

We are nearly done. Place a 3.5-cm piece of the thin double-sided tape on the large tab under the eyes and carefully insert this into the large tab space between the two limit switches on the robot base. Once you have it lined up and positioned how you like, make the positive connection between the taped tab and the base of the robot.

It was during this step that my motor base popped apart, and I had to add some extra tape to hold it in place. You can see in the image below how I reinforced the assembly. I also added a few pieces here and there to corner tabs that looked like they might come loose. This is standard, clear tape and NOT the double-sided stuff from the kit.

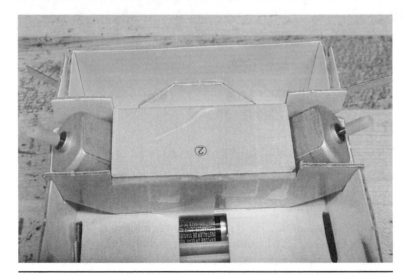

First connection of the robot top and reinforcement

Step Eighteen

To finish attaching the top of the robot, we line up the two tabs on the side and the one at the back and apply steady pressure until we seat the back tab. This tab is larger than its slot, which creates a great seal for the robot and allows us to pop the top later and change the batteries as needed.

Closing the top

All that is left is to put the paws on the limit switch arms. You will need two strips of the thick white double tape about 0.5 cm wide and 3 cm long. Once you have the paws attached with the tape, use some regular clear tape to wrap around the paws and arms in a couple of different directions to ensure they do not fall off when they hit objects. This is the part of the robot designed to hit the obstacles the system encounters, so you want to make sure they can stand up to some abuse.

Remember, anytime you have batteries in the robot and hit one of the limit switches, the motors will turn. This can make attaching the paws interesting if the robot is sitting on a table or workbench.

 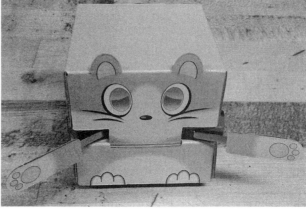

A couple of paws and ready to go!

Conclusion

You have just built a robot that will interact with its environment due to the installed sensors. This is a simple first step on the journey of understanding the robot, but a great place to begin experimentation. Now that you know how to reverse the motors with a limit switch, perhaps you want to do the same thing with the Box Robot you built previously, or maybe you will use this in one of your classroom labs. Every time you learn a way to control robot action, you increase your options in robot operation.

If this is an assigned lab, make sure you turn in the required paperwork. There are blank lab forms in the back of the activities manual to help with the reporting process if needed.

Spider Bot Lab

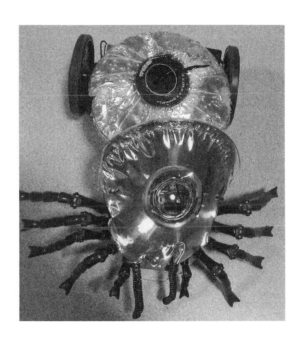

Materials

- Coaster Bot Challenge Kit from Jameco #2115741 (optional)
- Free-rolling wheel or coaster
- FTDI Basic Breakout (5v) from Solarbotics #50512
- CDs or DVDs (damaged or dead ones work great)
- Wire strippers
- Needle-nose pliers
- AAA batteries
- 9-V battery
- Soldering iron with solder
- Drill with drill bits
- Small Philips screwdriver
- Roll of electrical tap
- Glue gun or double-sided tape
- Fabricated or reused items

About the Lab

The Spider Bot was my entry in a *Make* magazine contest several years ago. While my robot did not place, it was a lot of fun to build, and I think this project is a great way to dive into the world of robot programming. The processor for this robot is the Ardweeny, which is a modified Arduino control designed to fit in small places. The Ardweeny and breadboard power supply come as solder together kits, so you get the added benefit of learning to solder while learning how the basic parts of an Arduino and power supply fit together.

The Spider Bot falls into a group of robots commonly called coaster robots. Coaster robots are systems that use CDs or similar for the rigid structure, with the internal workings being builders choice. At the time I built mine, I did not have many spare parts on hand, so I picked up the Coaster Bot Challenge Kit from Jameco #2115741 (if you go to www.jameco.com and type the part number into the search bar, it should bring up the kit). If you have a stockpile of parts and such, you may want to pick up the few things you are missing and save the cost of the kit. In the kit's description, it gives you the quantities and part numbers for everything in the kit, for those who want to source their own parts.

I built mine to look like a large spider, but you are free to build yours in whatever configuration you choose. The main point here is to learn how to hook all the components together and make a working system. I used limit switches for sensors, similar to the Obstacle Avoiding Robot Lab, but unlike that lab, the Spider Bot uses controller logic to reverse the motors instead of switching the wiring. This allows the programmer to control how the robot reacts when a sensor is triggered instead of reverting to normal operation as soon as the pressure is released.

The Parts

The following images show the various parts, besides the CDs and Ardweeny, I ordered to build my robot. If you have access to similar parts, there is a good chance you do not need to order a kit and can start building with what you have on hand and source the missing items. You should be able to substitute most Arduino controllers for the Ardweeny, as long as you make the proper pin connections.

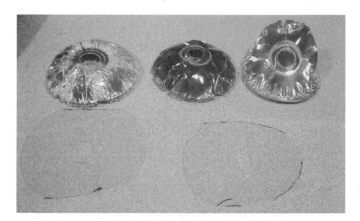
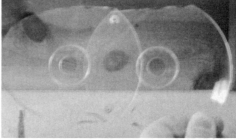

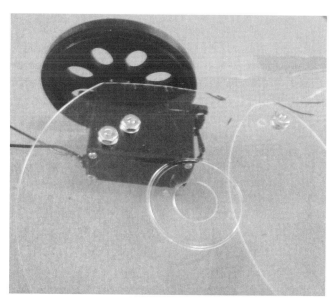

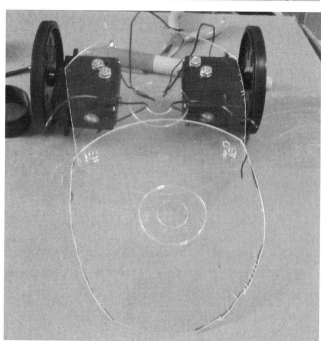

Jameco kit parts

Hacking the Servos

The two servos that came in the kit turned 180 degrees and stopped, but for the purpose of this robot, I needed continuous motion. To get around this, I had to modify the servos. You may be thinking to yourself, why not just use DC motors? The reason is speed and torque. Regular DC motors turn fast and have low torque, and for this robot we want slower speed and higher torque, which means gearing. If you remember back in Chapter 4 of the textbook, we talked about gear trains and how we sacrifice speed but gain torque. Since the wheels of the robot hook directly to the motors, we need some kind of gearing to gain torque, thus we are using geared servos instead of DC motors and gearboxes. If you have continuous rotation servos on hand, you can skip this step; otherwise, go to http://karldemuthrobotbuild.blogspot.com/2010/04 /servo-modification-for-continuous.html and follow the instructions for hacking a servomotor for continuous motion. Karl Demuth is the author of this work, and he does a great job of walking one through the process with many great pictures to help along the way.

If for some reason you cannot access this website, go to www.instructables.com, search "hacking a servo for continuous motion," and take a look at the several different instruction sets on how to perform this portion of the lab.

The Controller and Power Supply

The next thing I did was solder together the breadboard power supply and the Ardweeny. Each of these kits came with great instructions, so there is no reason for me to repeat the steps here. The power supply was a quick build and took less than a half hour, but the Ardweeny took a bit longer. To help with the process of putting it together, you can reference the following images to see how my Ardweeny came together. The assembly instructions will walk you through making sure you have components in the right spot and orientated properly. If you have an assembled Arduino controller you plan to use for the build, you can skip this step, though you may have to build or find a power supply for your breadboard if the Arduino you have in mind is not up to the task.

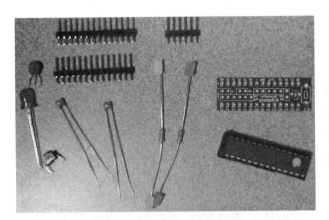
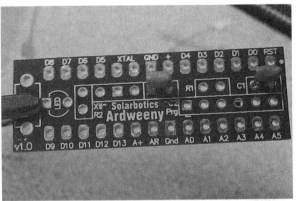
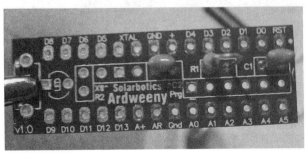

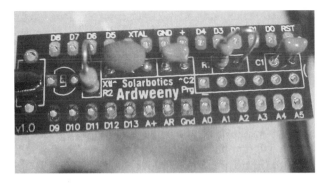

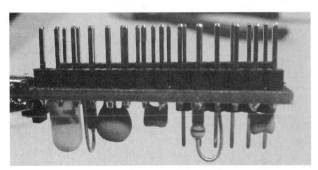

Starting the Build

Now that you have all the parts gathered, have continuous rotation servos, and have the controller and power supply ready, you can begin to turn this pile of parts into a robot.

I tried something fancy with the body of my robot, thermo forming the CDs. To do this, I used a thick ceramic bowl I bought from the store and my oven as a heat source. I had to play with the temperature because if it was too low, the CDs remained as they were; if it was too hot, they began to smoke. I think the next time I try something like this, I will use a heat gun instead of the oven due to the nature of the heat and the finer control I have over the heat source. If you are going to try something like this, I have a few cautions. First, make sure you have plenty of ventilation or fresh air. Second, do not use something to form the CDs to or around that you want to use for anything else. The bowl I used still has remnants of plastic stuck to it and is good for only this type of experimentation now. Third, hot plastic will burn you and the foil inside the CDs holds heat much longer than you think it should. I burned my fingers a couple of times handling the CDs too quickly after heat treatment.

For your robot, my recommendation is to use the CDs as they are and avoid trying anything fancy. The first time you build a robot like this, there are plenty of things to learn besides ways to change the shape of plastic. You may want to use one CD for the base and then add CDs as needed to complete the build, but the choice of shape is up to you. To keep with the pictures of this instruction, I will outline how I built the Spider Bot.

Body Parts

For the Spider Bot, I used two clear plastic CD covers from my burnable CDs and two of the three CDs I thermo formed for the body. Since my theme was a spider, I needed a long body with two segments rather than the traditional single CD body of most coaster bots. In the picture below, you will notice I trimmed the clear CD covers to create the long two-segment body of a spider.

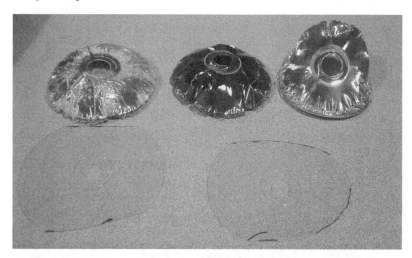

Body parts for the robot

Building the Base

The next step was to hook the two clear base pieces together so that I could mount the various parts to it. I used a drill to make the holes and then a couple of the nuts and bolts from the kit to hold everything in place. I discovered later that the clear CDs are a bit flimsy, and the base flexed due to this, but it did not impede operation at all. If you are using a single CD as the base, you can skip this step.

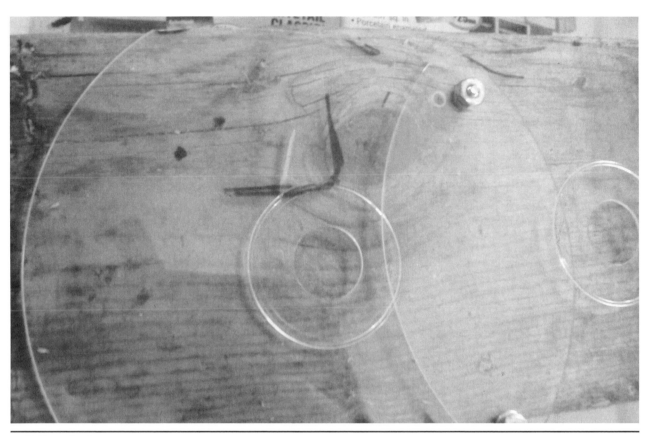

Building the base

Attaching the Servos

Now we are ready to put the servos on the base. Before you attach the servos, go ahead and add the wheels to the units so that you can make sure to position everything properly. When you are ready to attach the servos, you can use double-sided tape, hot glue, or bolt the servos in place. I wanted a solid build, so I opened my servo cases and made two holes in a safe location to attach my servos to the base, as none of the provided bolt holes matched up with the mounting orientation I had in mind. To be honest, I took a bit of a risk doing it this way because positioning the holes incorrectly could cause the bolt heads to interfere with the gearing of the servo, and any missed shavings could gum up the works. My recommendation is to go with the glue or double-sided tape on your first build to avoid potential issues. If you use glue, make sure to avoid getting any on the shaft of the servo.

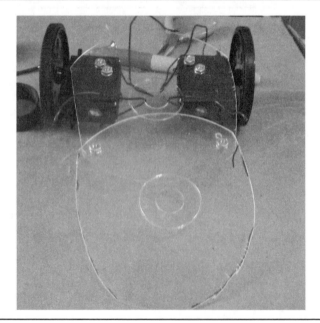

Attaching the servos

Attaching the Coaster

Since the robot has only two drive wheels, we need a third wheel to help everything move smoothly, especially given the design of this robot. To help with this, I removed the coaster from the same spider toy that I stole the spider legs off of (the spider legs are used later in the build). I drilled four holes in the plastic body around the wheel for mounting purposes and then drilled matching holes in the frame and used some of the mounting hardware from the kit to hold it in place. The coaster I used is a simple wheel that turns freely but stays straight. This added stability to the robot and helped prevent erratic movement during operation.

Adding the coaster wheel

Wiring in the Controller

Before we talk about wiring in the controller, take a few minutes to study the wiring diagram and pinout images below. This is the heart of the build and the key to making everything work, so you want to be sure you get it all wired together correctly.

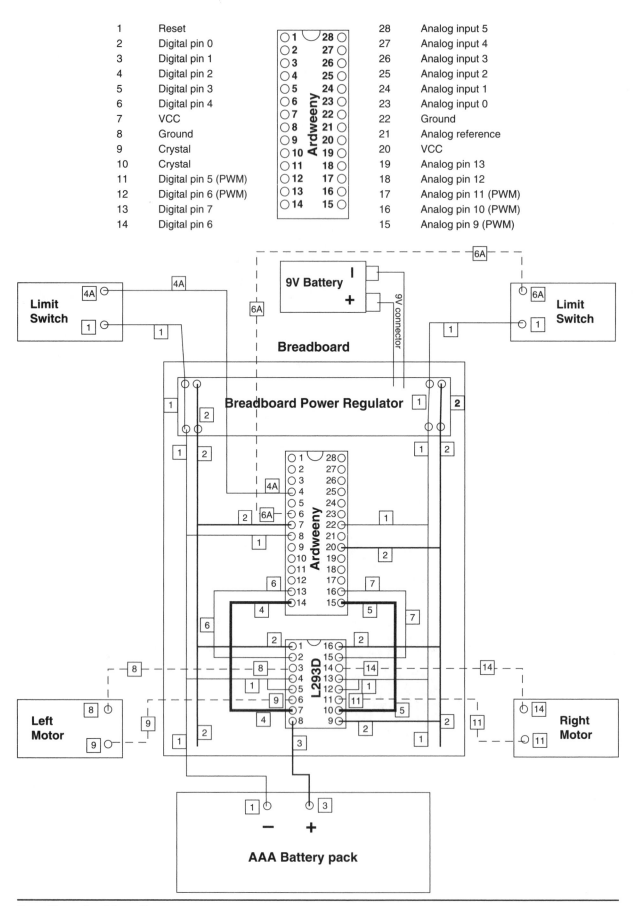

1	Reset	28	Analog input 5
2	Digital pin 0	27	Analog input 4
3	Digital pin 1	26	Analog input 3
4	Digital pin 2	25	Analog input 2
5	Digital pin 3	24	Analog input 1
6	Digital pin 4	23	Analog input 0
7	VCC	22	Ground
8	Ground	21	Analog reference
9	Crystal	20	VCC
10	Crystal	19	Analog pin 13
11	Digital pin 5 (PWM)	18	Analog pin 12
12	Digital pin 6 (PWM)	17	Analog pin 11 (PWM)
13	Digital pin 7	16	Analog pin 10 (PWM)
14	Digital pin 6	15	Analog pin 9 (PWM)

Breadboard wiring diagram

1	Motor 1 control	1 28	16	Analog input 5
2	Motor 1 logic pin 1	2 27	15	Motor 2 logic pin 1
3	Motor 1 terminal 1	3 26	14	Motor 2 terminal 1
4	Heat sink/ground	4 25	13	Heat sink/ground
5	Heat sink/ground	5 24	12	Heat sink/ground
6	Motor 1 terminal 2	6 23	11	Motor 2 terminal 2
7	Motor 1 logic pin 2	7 22	10	Motor 2 logic pin 2
8	Motor power supply	8 21	9	Motor 2 control

L293D

(Continued)

Refer to the following images to assist in wiring everything together. The images do not include the motor and limit switch connections, as we will add those later. In the pictures, you can see something connected on top of the Ardweeny, and this is the FTDI Basic Breakout (5v), which is used to communicate between the controller and the programming computer. If you used a regular Arduino controller, you likely do not need this item.

Pins 8 and 22 of the Ardweeny connect to the negative side of the breadboard power supply. Pins 7 and 20 connect to the positive side. Pin 13 of the Ardweeny connects to pin 2 of the L293D. Pin 14 of the Ardweeny connects to pin 7 of the L293D. Pin 16 of the Ardweeny connects to pin 15 of the L293D. Pin 15 of the Ardweeny connects to pin 10 of the L293D.

On the L293D, pins 4, 5, 12, and 13 all hook to the negative breadboard supply. Pins 1, 9, and 16 all connect to the positive breadboard supply or 5-V DC.

These connections give the Ardweeny and L293D the power they need to run and allow us to control the operation of the motors. Use the following pictures to assist in making your connections. If you are not familiar with breadboards, take a few minutes to figure out which sections connect together so that you can make the proper wiring connections. The components pictured are designed to fit on a breadboard, but they may be a bit of a tight fit. The Ardweeny and L293D both have a half moon cutout at one end to ensure proper pin orientation.

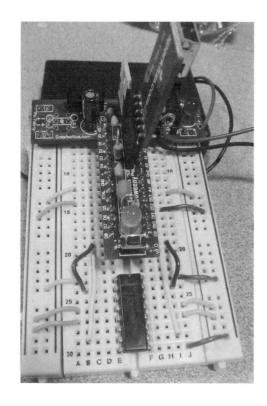

Triggering the Limit Switches

To interact with the robot's environment and trigger the limit switches, I used a set of spider legs stolen form a toy I had laying around. I anchored the base of each leg assembly in place and then left the other end free to move, as this is what would press against and trigger the limit switches when the robot hit an obstacle. I ended up adding some wire legs to the assembly because the plastic ones I had did not engage the switches as I had wanted (you will see a picture of this in the next step). For your robot, you can add long pieces of wire to your limit switch arms to act as triggers or use your imagination and creativity to find something that would add esthetic value to your robot while fitting the need of limit switch activation device.

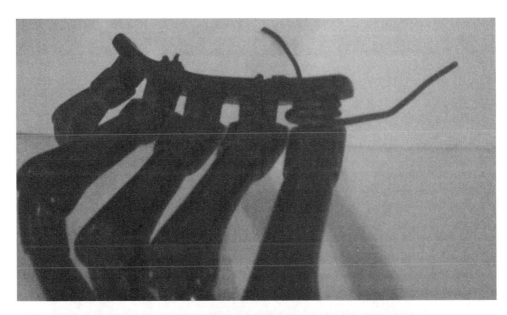

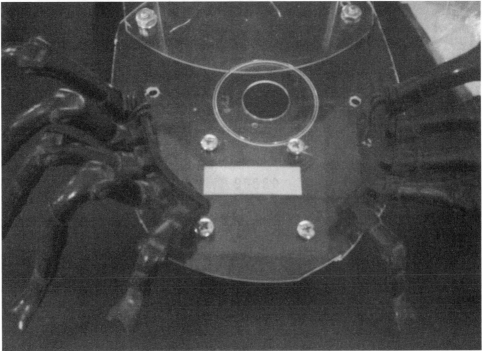

Attaching the legs

The Last of the Internal Components

Once I had the legs in place, I attached wires to the NO contacts and commons of the limit switches and then glued them in place where the legs could activate them. If you look closely, you can see I used the wire legs to make a connection between the 2 sets of spider legs. This added stability to the legs and helped with activating the limit switches. This setup worked decently, but I never was happy with the way it engaged the limit switches. When I get around to version two someday, I plan to find a different configuration to activate the limit switches. Perhaps I will tie them directly to the wire additions to the legs or use something similar. You may want to figure out a way to tie your switch activation device directly to the limit switch arm to ensure proper operation.

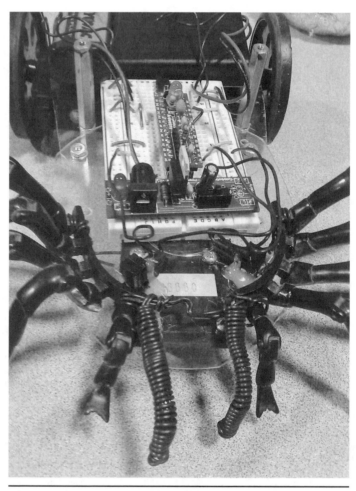

Adding in the limit switches

Next, I added the motor power supply, the 9-V battery for the breadboard power supply, and made the final connections to the breadboard. To hold the batteries in place, I used the spare wire from the motors to make a cradle they fit in. If you are worried about the power supplies working loose, some double-sided tape or a drop of hot glue should hold them firmly in place.

For the final electrical connections, I tied the commons of the limit switches to negative of the breadboard power supply and then the NO contacts to pins 4 and 6 of the Ardweeny with the left switch tied to pin 4. The AAA battery pack has the black lead tied to the negative breadboard power supply and the red lead tied to pin 8 of the L293D. This is the power supply for the two servomotors, and it runs through the L293D, which is designed to handle the load. Finally, the left motor is tied to pins 3 and 6 of the L293D, and the right motor is tied to pins 14 and 11.

The system is now ready for a program and testing!

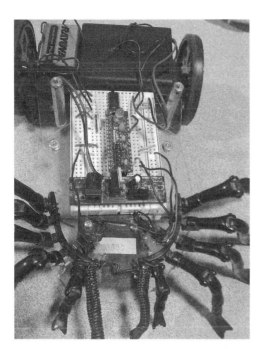

Final connections and placement

The Program

The Arduino programming software is a free download, and you will want to go online and track down the latest copy. You can find the latest version of the software at http://www.arduino.cc/en/Main/Software or by searching for Arduino software. Once you have the software downloaded, take the time to read the user help files, as they will assist you in figuring out how to use the software.

It took me several hours to piece together the program in my Spider Bot, and it is a cut and paste work from several program examples I found at various places on the Internet. If you have a background with C++-type programming, you will find the Arduino easy to program. If this is your first programming adventure, take the time to read through the beginners' tips and tricks you can find on the Arduino site and other places on the Internet.

There is a FAQ section and forum on the Arduino site that helped me with most of the questions I had. Below is a copy of the program that ran in my Spider Bot:

```
// These constants define which pins on the Ardweeny are connected to the pins on
// the motor controller. If your robot isn't moving in the direction you expect it
// to, you might need to swap these!
const unsigned char leftMotorA = 7;
const unsigned char leftMotorB = 8;

const unsigned char rightMotorA = 9;
const unsigned char rightMotorB = 10;

const unsigned char leftswitch = 2;
const unsigned char rightswitch = 4;

boolean leftswitchclosed;
boolean rightswitchclosed;

// This function is run first when the microcontroller is turned on
void setup() {
// Initialize the pins used to talk to the motors
pinMode(leftMotorA, OUTPUT);
pinMode(leftMotorB, OUTPUT);
pinMode(rightMotorA, OUTPUT);
pinMode(rightMotorB, OUTPUT);
pinMode(leftswitch, INPUT);
digitalWrite (leftswitch, HIGH);
pinMode(rightswitch, INPUT);
digitalWrite (rightswitch, HIGH);

// Writing LOW to a motor pin instructs the L293D to connect its output to ground.
digitalWrite(leftMotorA, LOW);
digitalWrite(leftMotorB, LOW);
digitalWrite(rightMotorA, LOW);
digitalWrite(rightMotorB, LOW);

}


// This function gets called repeatedly while the microcontroller is on.
void loop() {

if (digitalRead (leftswitch) == LOW)
{leftswitchclosed=true;}
else {leftswitchclosed=false;};

if (digitalRead (rightswitch) == LOW)
{rightswitchclosed=true;}
else {rightswitchclosed=false;};
```

```
// Turn both motors on, in the 'forward' direction
if (!rightswitchclosed && !leftswitchclosed)
{digitalWrite(leftMotorA, HIGH);
digitalWrite(leftMotorB, LOW);
digitalWrite(rightMotorA, HIGH);
digitalWrite(rightMotorB, LOW);}


// Turns robot right
else if (!rightswitchclosed && leftswitchclosed)
{digitalWrite(leftMotorA, HIGH);
digitalWrite(leftMotorB, LOW);
digitalWrite(rightMotorA, LOW);
digitalWrite(rightMotorB, HIGH);}


// Turn robot left
else if (rightswitchclosed && !leftswitchclosed)
{digitalWrite(leftMotorA, LOW);
digitalWrite(leftMotorB, HIGH);
digitalWrite(rightMotorA, HIGH);
digitalWrite(rightMotorB, LOW);}


// Robot goes in reverse
else
{digitalWrite(leftMotorA, LOW);
digitalWrite(leftMotorB, HIGH);
digitalWrite(rightMotorA, LOW);
digitalWrite(rightMotorB, HIGH);};

// Wait 1 second
delay(500);
}
```

Some Tips

My robot would turn whenever it hit an obstacle and then take off once more. During testing, I found mine had a tendency to drift to one side, and I believe this was due to the servo motor placement. If you experience a similar problem, you may want to adjust your motor placement. If you have trouble downloading your program to the controller, make sure the various power switches are on and that you have the power LED on the breadboard supply; without this, your Ardweeny has no juice. The help files on the Arduino site will likely get you past the problem, but you can always post on the forum for help from the Arduino community.

On the Ardweeny pins, HIGH means to send out voltage or check for signal present; LOW means connect the pin to ground. By switching the high and low status of the Ardweeny pins tied to the motor logic pins on the L293D, we reverse the rotation of the motors. Without 5 V to the motor control pins on the L293D, there is no motor rotation regardless of what we do to the motor logic pins. Without 5 V to the logic power supply, the L293D will simply not work. If you need to troubleshoot the circuit logic deeper than this, I recommend you search for specific information on the L293D and Arduino chips.

One last tip—I tried to download my program from a school computer in the lab and never could get the program to download, but it downloaded fine at home. This leads me to believe that school firewalls may cause problems with program downloading, which is something to keep in mind.

Once you have a successful test of the system you are ready to put on the final touches.

The Finishing Touches

To finish off my Spider Bot, I carefully drilled holes in the two thermo formed CDs I chose and bolted them in place. I used the standoff posts that came with the robot kit to make these connections and used the odd-shaped CD for the headpiece and the round one for the abdomen. You can see some of the inner working through the center of the CD head, and I liked the overall effect. You can look at the pictures below to see how the finished robot turned out. I later discovered that thermo forming CDs makes them brittle, as the Spider Bot shell has started to crack and break over time.

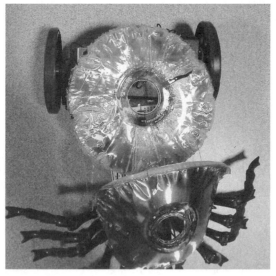

The finished robot

Conclusion

You now know the basic steps of building your own programmable, autonomous robot. As you learn more about sensors, programming, and robotics, I encourage you to revisit this lab and find ways to integrate what you have learned into the system. The options for functionality and design are limited only by your ingenuity and knowledge.

If this is an assigned lab, make sure you turn in the required paperwork. There are blank lab forms in the back of the activities manual to help with the reporting process if needed.

SECTION 3

General Lab Forms

About This Section

This section of the activities manual has the general lab form designed to help students organize the information from labs in the classroom. This form is designed to help with the lab reporting process and DOES NOT take the place of any forms given to you by your instructor. Make sure you answer each section as best you can and if you run out of room in any of the sections, you can attach another piece of paper to the form. Make sure you put the name of everyone in your lab group on the form so that your instructor can properly assign credit.

For grading guidelines on labs and the points possible, ask your instructor. Make sure to follow all the safety rules of your classroom and to keep your wits about you as you complete labs. Labs are often the more exciting part of a robotics class, but they are also the most dangerous. Remember the three Rs of robotics: Robots Require Respect. Before you complete your first lab of the class, I recommend you review Chapter 2 of the textbook on safety.

I hope you enjoy your lab exercises and learn a great deal about robotics. Lab activities are most students' favorite part of the robotics courses, and I imagine many of you reading this feel the same way. When it comes to labs, it is key to remember the more work and effort you put into the activity, the more you will learn about robotics.

General Lab Form

Lab Description

In the space below, describe the purpose of the lab and the equipment involved.

_____
_____
_____
_____
_____
_____
_____
_____
_____
_____

Lab Execution

In the space provided, detail the steps you took to perform the lab. Make sure to include any troubleshooting steps performed.

_____
_____
_____
_____
_____
_____
_____
_____
_____
_____
_____

Observations

Record your observations about the system's performance here, including both the expected and unexpected.

_____

_____

_____

_____

_____

_____

_____

_____

_____

_____

Conclusions

What conclusions or statements can you make about the robot based on your observations and any data gathered during the course of the lab?

_____

_____

_____

_____

_____

_____

_____

_____

_____

_____

General Lab Form

Lab Description

In the space below, describe the purpose of the lab and the equipment involved.

_____
_____
_____
_____
_____
_____
_____
_____
_____

Lab Execution

In the space provided, detail the steps you took to perform the lab. Make sure to include any troubleshooting steps performed.

_____
_____
_____
_____
_____
_____
_____
_____
_____
_____

Observations

Record your observations about the system's performance here, including both the expected and unexpected.

_____

_____

_____

_____

_____

_____

_____

_____

_____

Conclusions

What conclusions or statements can you make about the robot based on your observations and any data gathered during the course of the lab?

_____

_____

_____

_____

_____

_____

_____

_____

_____

General Lab Form

Lab Description

In the space below, describe the purpose of the lab and the equipment involved.

_____
_____
_____
_____
_____
_____
_____
_____
_____
_____

Lab Execution

In the space provided, detail the steps you took to perform the lab. Make sure to include any troubleshooting steps performed.

_____
_____
_____
_____
_____
_____
_____
_____
_____
_____

Observations

Record your observations about the system's performance here, including both the expected and unexpected.

_____

_____

_____

_____

_____

_____

_____

_____

Conclusions

What conclusions or statements can you make about the robot based on your observations and any data gathered during the course of the lab?

_____

_____

_____

_____

_____

_____

_____

_____

General Lab Form

Lab Description

In the space below, describe the purpose of the lab and the equipment involved.

_____

_____

_____

_____

_____

_____

_____

_____

_____

_____

Lab Execution

In the space provided, detail the steps you took to perform the lab. Make sure to include any troubleshooting steps performed.

_____

_____

_____

_____

_____

_____

_____

_____

_____

_____

Observations

Record your observations about the system's performance here, including both the expected and unexpected.

_____

_____

_____

_____

_____

_____

_____

_____

_____

_____

Conclusions

What conclusions or statements can you make about the robot based on your observations and any data gathered during the course of the lab?

_____

_____

_____

_____

_____

_____

_____

_____

_____

General Lab Form

Lab Description

In the space below, describe the purpose of the lab and the equipment involved.

_____
_____
_____
_____
_____
_____
_____
_____
_____

Lab Execution

In the space provided, detail the steps you took to perform the lab. Make sure to include any troubleshooting steps performed.

_____
_____
_____
_____
_____
_____
_____
_____
_____
_____

Observations

Record your observations about the system's performance here, including both the expected and unexpected.

_____

_____

_____

_____

_____

_____

_____

_____

_____

Conclusions

What conclusions or statements can you make about the robot based on your observations and any data gathered during the course of the lab?

_____

_____

_____

_____

_____

_____

_____

_____

_____

_____

General Lab Form

Lab Description

In the space below, describe the purpose of the lab and the equipment involved.

_____
_____
_____
_____
_____
_____
_____
_____
_____
_____

Lab Execution

In the space provided, detail the steps you took to perform the lab. Make sure to include any troubleshooting steps performed.

_____
_____
_____
_____
_____
_____
_____
_____
_____
_____

Observations

Record your observations about the system's performance here, including both the expected and unexpected.

_____

_____

_____

_____

_____

_____

_____

_____

_____

Conclusions

What conclusions or statements can you make about the robot based on your observations and any data gathered during the course of the lab?

_____

_____

_____

_____

_____

_____

_____

_____

_____

General Lab Form

Lab Description

In the space below, describe the purpose of the lab and the equipment involved.

_____

_____

_____

_____

_____

_____

_____

_____

_____

_____

Lab Execution

In the space provided, detail the steps you took to perform the lab. Make sure to include any troubleshooting steps performed.

_____

_____

_____

_____

_____

_____

_____

_____

_____

_____

Observations

Record your observations about the system's performance here, including both the expected and unexpected.

_____

_____

_____

_____

_____

_____

_____

_____

_____

_____

Conclusions

What conclusions or statements can you make about the robot based on your observations and any data gathered during the course of the lab?

_____

_____

_____

_____

_____

_____

_____

_____

_____

_____

General Lab Form

Lab Description

In the space below, describe the purpose of the lab and the equipment involved.

_____

_____

_____

_____

_____

_____

_____

_____

_____

_____

Lab Execution

In the space provided, detail the steps you took to perform the lab. Make sure to include any troubleshooting steps performed.

_____

_____

_____

_____

_____

_____

_____

_____

_____

_____

Observations

Record your observations about the system's performance here, including both the expected and unexpected.

_____

_____

_____

_____

_____

_____

_____

_____

_____

Conclusions

What conclusions or statements can you make about the robot based on your observations and any data gathered during the course of the lab?

_____

_____

_____

_____

_____

_____

_____

_____

_____

General Lab Form

Lab Description

In the space below, describe the purpose of the lab and the equipment involved.

_____
_____
_____
_____
_____
_____
_____
_____
_____
_____

Lab Execution

In the space provided, detail the steps you took to perform the lab. Make sure to include any troubleshooting steps performed.

_____
_____
_____
_____
_____
_____
_____
_____
_____
_____

Observations

Record your observations about the system's performance here, including both the expected and unexpected.

_____

_____

_____

_____

_____

_____

_____

_____

_____

_____

Conclusions

What conclusions or statements can you make about the robot based on your observations and any data gathered during the course of the lab?

_____

_____

_____

_____

_____

_____

_____

_____

_____

_____

General Lab Form

Lab Description

In the space below, describe the purpose of the lab and the equipment involved.

_____
_____
_____
_____
_____
_____
_____
_____
_____
_____

Lab Execution

In the space provided, detail the steps you took to perform the lab. Make sure to include any troubleshooting steps performed.

_____
_____
_____
_____
_____
_____
_____
_____
_____
_____
_____
_____

Observations

Record your observations about the system's performance here, including both the expected and unexpected.

_____

_____

_____

_____

_____

_____

_____

_____

_____

Conclusions

What conclusions or statements can you make about the robot based on your observations and any data gathered during the course of the lab?

_____

_____

_____

_____

_____

_____

_____

_____

_____

General Lab Form

Lab Description

In the space below, describe the purpose of the lab and the equipment involved.

_____

_____

_____

_____

_____

_____

_____

_____

_____

_____

Lab Execution

In the space provided, detail the steps you took to perform the lab. Make sure to include any troubleshooting steps performed.

_____

_____

_____

_____

_____

_____

_____

_____

_____

_____

Observations

Record your observations about the system's performance here, including both the expected and unexpected.

_____

_____

_____

_____

_____

_____

_____

_____

Conclusions

What conclusions or statements can you make about the robot based on your observations and any data gathered during the course of the lab?

_____

_____

_____

_____

_____

_____

_____

_____

General Lab Form

Lab Description

In the space below, describe the purpose of the lab and the equipment involved.

_____

_____

_____

_____

_____

_____

_____

_____

_____

_____

_____

Lab Execution

In the space provided, detail the steps you took to perform the lab. Make sure to include any troubleshooting steps performed.

_____

_____

_____

_____

_____

_____

_____

_____

_____

_____

_____

Observations

Record your observations about the system's performance here, including both the expected and unexpected.

_____

_____

_____

_____

_____

_____

_____

_____

_____

Conclusions

What conclusions or statements can you make about the robot based on your observations and any data gathered during the course of the lab?

_____

_____

_____

_____

_____

_____

_____

_____

General Lab Form

Lab Description

In the space below, describe the purpose of the lab and the equipment involved.

_____

_____

_____

_____

_____

_____

_____

_____

_____

_____

Lab Execution

In the space provided, detail the steps you took to perform the lab. Make sure to include any troubleshooting steps performed.

_____

_____

_____

_____

_____

_____

_____

_____

_____

_____

Observations

Record your observations about the system's performance here, including both the expected and unexpected.

Conclusions

What conclusions or statements can you make about the robot based on your observations and any data gathered during the course of the lab?

General Lab Form

Lab Description

In the space below, describe the purpose of the lab and the equipment involved.

_____
_____
_____
_____
_____
_____
_____
_____
_____
_____
_____

Lab Execution

In the space provided, detail the steps you took to perform the lab. Make sure to include any troubleshooting steps performed.

_____
_____
_____
_____
_____
_____
_____
_____
_____
_____
_____

Observations

Record your observations about the system's performance here, including both the expected and unexpected.

_____

_____

_____

_____

_____

_____

_____

_____

_____

Conclusions

What conclusions or statements can you make about the robot based on your observations and any data gathered during the course of the lab?

_____

_____

_____

_____

_____

_____

_____

_____

_____

_____

_____

General Lab Form

Lab Description

In the space below, describe the purpose of the lab and the equipment involved.

_____
_____
_____
_____
_____
_____
_____
_____
_____
_____

Lab Execution

In the space provided, detail the steps you took to perform the lab. Make sure to include any troubleshooting steps performed.

_____
_____
_____
_____
_____
_____
_____
_____
_____
_____
_____
_____

Observations

Record your observations about the system's performance here, including both the expected and unexpected.

_____
_____
_____
_____
_____
_____
_____
_____
_____
_____

Conclusions

What conclusions or statements can you make about the robot based on your observations and any data gathered during the course of the lab?

_____
_____
_____
_____
_____
_____
_____
_____
_____
_____